Essentials of Statistics

The Institute of Marketing

Marketing means Business

The Institute of Marketing was founded in 1911. It is now the largest and most successful marketing management organisation in Europe with over 20,000 members and 16,000 students throughout the world. The Institute is a democratic organisation and is run for the members by the members with the assistance of a permanent staff headed by the Director General. The Headquarters of the Institute are at Moor Hall, Cookham, near Maidenhead, in Berkshire.

Objectives: The objectives of the Institute are to develop knowledge about marketing, to provide services for members and registered students and to make the principles and practices of marketing more widely known and used throughout industry and commerce.

Range of activities: The Institute's activities are divided into four main areas:
   Membership and membership activities
   Corporate activities
   Marketing education
   Marketing training

Essentials of Statistics in Marketing

C. S. GREENSTED,

B.Sc., M.Sc., Lecturer, Department of Accountancy and Finance,
University of Strathclyde

A. K. S. JARDINE,

B.Sc., M.Sc., Ph.D., C.Eng., M.I.Mech.E., M.I.Prod.E.
Professor-in-Charge, Engineering and Management,
Royal Military College of Canada

J. D. MACFARLANE,

B.Sc., Senior Lecturer and
Director of the Health Services Operational Research Unit,
University of Strathclyde

Published on behalf of

the INSTITUTE OF MARKETING
and the CAM FOUNDATION

HEINEMANN PROFESSIONAL PUBLISHING

William Heinemann Professional Publishing Ltd.
22 Bedford Square, London WC1B 3HH

LONDON   MELBOURNE   AUCKLAND

First published 1974
Second edition 1978
Reprinted 1980, 1982, 1985, 1986, 1988
ISBN 0 434 90887 8

Printed in Great Britain by
LR Printing Services Ltd., Crawley, W. Sussex

Foreword

This book can be counted upon to help readers in the understanding of numerical data and, in consequence, to improve the quality of their decision-making. It has been written, after detailed research, by a team of authors from the University of Strathclyde, which has long sought to make a practical contribution to marketing studies.

The requirement of a marketing career, and of the IM Diploma examinations catering for this, inevitably attract a range of candidates who do not share a common level of numeracy. Taking account of this fact, the authors have wisely prepared their material in such a way that it will be of equal use and appeal to students lacking statistical understanding as well as to those more familiar with mathematical notation and explanation.

I am very pleased, on behalf of the Institute of Marketing, to recommend this book to all IM members and, more particularly, to our students as essential reading for the IM diploma. I believe it will also be of value to students taking courses with a marketing content at universities and other colleges.

PETER B. BLOOD
Director-General
Institute of Marketing

Preface

Marketing and sales management usually have large quantities of information available to them in the form of sales returns, stock positions, accounting information, and market research data. The information is often presented as a mass of numbers in which there is no logical form and from which it is difficult to draw conclusions as to the true state of the market and the business. However, management decisions are often ostensibly based on this mass of fuzzy information.

This book has stemmed both from short courses given to practising marketing and sales management and from postgraduate courses given to Master of Business Administration students. It aims at showing such people methods by which the mass of data may sensibly be reduced in volume and hence how reasonable inferences may be drawn from it. It is then possible to make better decisions based on the kernel of the information contained in the original data.

We have tried to achieve the above aim in two ways. Firstly, we hope that the reader will understand sufficient of the methods to be able to carry out relatively simple analyses of data himself. Secondly, we hope that he will be able to recognize situations in future in which professional statistical analysis would yield useful information. We hope that he will then have sufficient knowledge of statistical concepts to be able to communicate with the statistician, understand his results, and perhaps question doubtful assumptions. In this second context some more esoteric concepts such as Bayesian statistics have been introduced.

An introduction to mathematical notation is given in Chapter 1 and we have tried to keep the level of mathematics in the text as low as possible. Some of the logic has been amplified mathematically in appendixes to the book but the non-mathematical reader may omit these appendixes without losing the gist of the argument. Exercises are given at the end of most chapters so that the reader may test his understanding of the text. Answers and a list of references for the reader who wishes to pursue the subject are given at the end of the book.

We are indebted to the literary executor of the late Sir Ronald A. Fisher, F.R.S., to Dr Frank Yates, F.R.S., and to Oliver & Boyd, Edinburgh, for permission to reprint Tables from their book *Statistical Tables for Biological, Agricultural and Medical Research*. We are also indebted to the Institute of Marketing (IM), the Institute of Statisticians (IOS), and the University of Strathclyde for permission to use some of their past examination questions as exercises. The source of individual exercises is acknowledged where relevant. Any acknowledgment prefixed with 'from' indicates that only part of a question is used. It should also be noted that some questions have been modified to bring them into line with both decimalization of money and metrication of units.

We should like to thank Drs Paul Davis of the University of Birmingham and Tony Christer of the University of Strathclyde for their helpful suggestions on modifications to an earlier draft manuscript.

C. S. GREENSTED
A. K. S. JARDINE
J. D. MACFARLANE

PREFACE TO THE SECOND EDITION

We are grateful to Dr C. Chatfield (University of Bath) for pointing out a number of errors which we have rectified in this edition.

C. S. GREENSTED
A. K. S. JARDINE
J. D. MACFARLANE

Contents

1. Why Statistics?

1.1 STATISTICS AND MANAGEMENT

Management decision making almost always involves a gamble. The principal reason for this is that in real life situations it is rarely possible to have *complete* information about the factors which affect the decision outcome and hence it is necessary to *infer* the *probable* consequences instead of *deducing* the *necessary* consequences of a decision. This difficulty is not peculiar to management situations. Historically the development of statistical methods was due principally to the recognition of this problem as scientists moved out of the carefully 'controlled' observational environment of the laboratory into the comparatively 'uncontrolled' environment of the real world. With a little imagination an analogy can be drawn between the problems of estimating the yield from a new type of wheat and estimating the market for a new product. Though one may be fertilized and the other advertised the ultimate outcome in each case is affected by a variety of uncontrolled or uncontrollable factors (of which the most tempting comparison is between the fickleness of weather and of public reaction!).

Given that management decisions must be taken in the light of incomplete information, what is the role of the statistician and his skills in this process? The main point to emphasize is that the use of statistical methods does not imply that by some magic the uncertainty associated with the incompleteness of the information is removed. What the statistician can do is recognize explicitly the magnitude of this uncertainty and incorporate this into the answers he gives to particular questions. To revive the gambling metaphor the statistician's role can often be regarded as that of 'getting the odds right'. This in turn implies that the answers which can be given to questions requiring statistical treatment of data are seldom straightforward 'yes' or 'no' answers but are usually qualified in some way. Management frequently complains of this lack of straightforwardness. This is not a consequence of deficiencies in the statistician's methods but is rather a natural attribute of the types of problem with which management is called to deal. It is only a slight exaggeration to claim that if a decision situation is so well defined as to permit an unqualified 'straightforward' answer then it is unlikely to call for any highly developed managerial skill.

1.1.1 STATISTICAL QUESTIONS AND ANSWERS

Some of the particular features of answers to statistical questions can be illustrated using the following simple sets of numbers. The data is to be understood as the weekly values of orders taken by two salesmen, Adams and Carter, over a 3 year period. It is assumed that the potential sales in each area are the same for each salesman in order to simplify the discussion. In practice this assumption would require to be validated.

SALES (in £100s)

Week ending	Adams	Carter
7/1/X1	25	17
14/1/X1	20	18
21/1/X1	18	23
28/1/X1	22	20
⋮	⋮	⋮
3/12/X3	34	28
10/12/X3	30	33
17/12/X3	32	39
24/12/X3	28	29
31/12/X3	14	17

Data of this kind might be considered relevant to answering questions of the following kind:

1. Is Adams a 'better' salesman than Carter or vice versa?
2. What is the probability of their combined sales exceeding £6,000/week?
3. Is there evidence of a trend towards increasing sales?
4. Is there evidence of a seasonal effect on sales and if so how great is it?

It is important to note that though these numbers have been placed in a particular context the basic characteristic of the set is that for each of a series of successive time periods a pair of numbers has been listed which are ostensibly comparable. Given this situation there exists, regardless of context, a set of general questions completely analogous to those given above. Thus the above numbers might refer to rainfall in two different localities, to kilometres run by vans from two different centres or to tonnes of iron produced by two blast furnaces. In any of these cases questions can be posed in the general form:

1. Is 'group A' greater than 'group B'?
2. What is the probability of occurrence of a combined value greater than 6,000?
3. Is there evidence of a trend towards higher values?
    etc.

Having underlined the fact that the raw data for statistical analysis can be considered, for most purposes, independently of the units of measurement involved, it is now important to consider some of the sub-problems which arise in trying to formulate answers to questions of the kind posed above. Consider Question 1—'Is Adams a "better" salesman than Carter?' Each salesman is represented by a group of 156 numbers. If Adams in every week had greater sales than Carter a case would exist for saying that he was the better salesman. Since this is not so the comparison must be based on a procedure which takes account of both the week-by-week differences between the two salesmen's returns *and* the pattern of variability of these differences. In its most general form this procedure requires that the information contained in these groups of numbers should be condensed into some synoptic or 'shorthand' description of the groups. In the next chapter a case is made for usually using the *arithmetic mean* and *standard deviation* of each group as the most appropriate measures. The question as posed invites one to arrive at a solution in which a statement is made that one salesman is 'better' than the other. It is known by inspection of the data that such a statement will not be true for every week recorded. What interpretation should be placed on such a statement? In Chapter 3 it is shown that meaningful statements can be made about the *probability* of one salesman's orders being greater than the other's. This term occurs in many statistically based answers to questions. The term, of course, is familiar but it is difficult to formulate a *formal* definition readily applicable to every situation. A discussion of some aspects of probability is offered in Chapter 3 but the reader is advised to fix in his mind the point that probability is concerned with quantifying one's uncertainty of or inadequacy of information about a certain situation. Probabilities can usually be interpreted as estimating the proportion of times a certain result is to be expected in repeated trials. The points just made have been based on only one of the suggested questions. Consideration of any of the other questions would have given rise to similar sets of sub-problems. This book is about such problems and the ways in which answers to them may be obtained.

Although statistical methods depend upon statistical theory the mathematical arguments have been eliminated wherever possible in this book. This does not mean that it has been possible to eliminate the use of mathematical *symbols* and *expressions*. The non-mathematical reader is asked to have courage! The expressions given, no matter how formidable they may seem at first glance, are simply statements about the way in which numbers are to be manipulated. They are instructions to carry out the simple operations of addition, subtraction, multiplication, and division, with only occasional excursions into slightly more complex operations such as determining the square root of a quantity (which can be found from tables in a few seconds). The apparent complexity arises because the statistician finds it convenient to express these manipulation instructions in 'shorthand' form. These 'shorthand' conventions are easily acquired and the most useful of them are explained in the following section.

1.2 BASIC MATHEMATICAL NOTATION

1.2.1 VARIABLES

It is simpler to analyse the sales returns of the two salesmen above by working in terms of mathematical symbols. One could denote the monthly sales returns of a salesman by the symbol y. For a particular month, y takes a particular numerical value and indeed takes different numerical values in different months, e.g.

in month 1, $y = 85$
„ „ 2, $y = 92$
$\vdots$
„ „ 12, $y = 104$

The values that y can take vary from month to month and therefore y is known as a *variable*. If the values of y vary 'randomly' in a way as yet undefined, y is sometimes referred to as a *variate* in statistical terminology. The choice of letter to assign to a variable is arbitrary although by convention the letters at the end of the alphabet are usually used. In the above situation one might represent 'the number of customers called on per month by a salesman' by the variable x.

For the purpose of this book, two types of variable may be distinguished.

(*a*) *Discrete variable.* A *discrete* variable may only take exact values, usually whole numbers (*integers*). The values are obtained by a counting process.

The number of calls made per month must be a whole number since a salesman cannot make a quarter of a call. The variable x above is therefore discrete. If sales are measured in units of number of cars, then sales return is a discrete variable.

An example of discrete values which are not integers is the decimal currency system. The values £1·00, £1·50, £2·75, etc., exist and are exact but the values £1·002, £2·491, etc., do not exist.

(*b*) *Continuous variable.* A *continuous* variable such as rainfall or time may theoretically take an infinite number of values between two fixed points on the scale.

The sales returns on a chemical product may be in units of tonnes and it is conceivable that a salesman could sell any fraction of a tonne. He could sell 1 tonne, 1 kilogram, 1 gram, or a fraction of a gram. He could theoretically sell any of an infinite number of different quantities between 0 and 1 tonne and is only limited by the accuracy of his weighing equipment. Although sales return is strictly a discrete variable since its value is obtained by counting the number of tonnes and fractions thereof sold, it is often convenient to treat it as a continuous variable. In general those variables whose values are obtained by taking 'fine' measurements are treated as continuous variables.

1.2.2 CONSTANTS

It seems reasonable to expect that the monthly number of customers called on would affect the monthly sales return and thus that a relationship would exist between the variables y and x, i.e. that y depends on x. This relationship may be directly proportional and take the mathematical form:

$$y = a + bx$$

i.e. $y = a + (b$ multiplied by $x)$.

This implies that for every increase in the value of x of one unit (i.e. one more call per month), monthly sales increase by b units. If no calls are made (i.e. $x = 0$), then sales will still reach some base level of a units. a and b represent fixed numbers which specify the relationship and are known as the *constants* of the equation or relationship. The precise relationship may vary from one salesman to the other and hence a and b may take different values for each salesman, e.g.

for Adams, sales return = 30 + 3 times (number of calls)
or $\qquad y = 30 + 3 . x$ (where . means 'multiplied by')
Here $a = 30, b = 3$
for Carter, sales return = 25 + 3·5 times (number of calls)
or $\qquad y = 25 + 3·5 . x$
Here $a = 25, b = 3·5$

By convention, constants are often denoted by the letters at the beginning of the alphabet. Some numbers appear very frequently in mathematics and have had letters permanently assigned to them. These special constants are often denoted by Greek characters—such as π (pi) which is approximately equal to 3·142. π is the ratio of the circumference to the diameter of a circle and is constant for all circles. The constant e will occur later in this book and has a value of 2·718 to three decimal places. It is closely related to the mathematical definition of logarithms.

1.2.3 SUPERSCRIPTS OR POWERS

It may be that the relationship between the two variables is not of the simple form shown above but that sales may depend on the *square* of the number of calls, i.e.

$$y = a + bx^2 = a + b . x . x$$

The '2' above the x is known as a *superscript* (or *power*) and we say it as 'x squared'. The superscript may be either a whole number or a fraction. In particular:

$x^{\frac{1}{2}} = $ square root of $x = \sqrt{x}$
$x^{-a} = 1/x^a$

Certain rules apply to superscripts:

1. Multiplication rule: $y^a \times y^b = y^{a+b}$
2. Division rule: $y^a \div y^b = y^{a-b}$
3. Power rule: $(y^a)^b = y^{ab}$
4. Special case: $y^0 = 1$

1.2.4 SUBSCRIPTS

In the above example, y has been defined as the monthly sales return without specifying which month or which salesman it concerns. When talking in general terms about sales this definition is sufficient. However it may be necessary to consider the sales in a particular month and this may be specified mathematically by the use of *subscripts*.

Subscripts take the form y_i where i is the subscript and, in the above example, refers to the ith month, e.g.

$y_1 =$ sales in month 1
$y_2 =$　,,　　　,,　　　,, 2
$\vdots$
$y_{12} =$　,,　　　,,　　　,, 12

Thus i can take values from 1 to 12 corresponding to the month numbers. Subscripts usually denote different values of the same variable. The letters in the middle of the alphabet are generally used as subscripts (unless using actual numbers) and may be called integer counters.

It is possible to put more than one subscript on a variable. The subscript i above refers to the month in which the sales return was made but there has been no differentiation between one salesman and the other. The subscript j may be introduced which could have a value 1 for Adams and 2 for Carter, i.e.

y_{ij} refers to sales return in month i made by salesman j and is a general value of the data, e.g.
$y_{6,2} = 74$ means that 74 sales were made in the 6th month by Carter.

1.2.5 INEQUALITY SIGNS

In assessing the two salesmen, the sales manager may wish to compare the sales returns for each month. He then sees whether Adams has sold less than, more than, or the same as Carter in each month. Less than or greater than are denoted mathematically thus:

$<$ means less than;
$\leqslant$ means less than or equal to;
$>$ means greater than;
$\geqslant$ means greater than or equal to.

The sales manager's comparison may be put symbolically as:

Is $y_{i1} < y_{i2}$ for all $i = 1, 2, \ldots 12$?

i.e. Are Adams' sales less than Carter's for each month 1 to 12?

1.2.6 SUMMATION

The sales manager may wish to know the total sales for each salesman in the year, for which knowledge he would add up the sales in each month over the year, i.e.

for Adams, the addition may be written algebraically as: $y_{1,1} + y_{2,1} + y_{3,1} + y_{4,1} + \ldots + y_{12,1}$

However it is a lengthy process writing out the addition as a series of terms and it may be written in shorthand using the symbol Σ (Greek capital sigma or S) which means 'the sum of'. Limits are put on the summation sign which show over which terms the summation is to be taken, e.g.

$$\sum_{i=1}^{12} y_{i1} = y_{1,1} + y_{2,1} + y_{3,1} + \ldots + y_{12,1}$$

In general the symbol $\sum_{i=1}^{n}$ means that i is replaced wherever it appears after the summation symbol by 1, then by 2, and so on up to n, and then the terms are added up.

In cases where the summation is to be taken over all possible values of the subscript (i) the limits are sometimes omitted, i.e.

$\sum_{i} y_{i1}$ is merely summing Adam's monthly sales

In some other cases only the summation sign is given, e.g. Σy_{i1}. If the sales manager required to know the total sales in the year, this could be written as a double summation. Thus:

$$\sum_{j=1}^{2} \sum_{i=1}^{12} y_{ij}$$

means the sum of all the ys taken over the i range for $j = 1$ plus the sum of the ys over the i range for $j = 2$, i.e.

$$\sum_{j} \sum_{i} y_{ij} = y_{1,1} + y_{2,1} + \ldots + y_{12,1} + y_{1,2} + y_{2,2} + \ldots + y_{12,2}$$

$\sum_{i=1}^{12} \sum_{j=1}^{2} y_{ij}$ means the sum of ys taken over both the js (i.e. 1 and 2) for

$i = 1$ plus the sum of the ys taken over both js for $i = 2$ and so on up to the sum of the ys taken over both js for $i = 12$, i.e.

$$\sum_i \sum_j y_{ij} = y_{1,1} + y_{1,2}$$
$$+ y_{2,1} + y_{2,2}$$
$$+ y_{3,1} + y_{3,2}$$
$$+$$
$$\vdots$$
$$+ y_{12,1} + y_{12,2}$$

Thus it can be seen that:

$$\sum_{j=1}^{2} \sum_{i=1}^{12} y_{ij} = \sum_{i=1}^{12} \sum_{j=1}^{2} y_{ij} = G$$

This equivalence is demonstrated in Table 1.1.

TABLE 1.1

	SALESMAN		
Month i	Adams $j = 1$	Carter $j = 2$	Total R_i
1	$y_{1,1}$	$y_{1,2}$	$\sum_j y_{1j}$
2	$y_{2,1}$	$y_{2,2}$	$\sum_j y_{2j}$
3	$y_{3,1}$	$y_{3,2}$	$\sum_j y_{3j}$
4	$y_{4,1}$	$y_{4,2}$	$\sum_j y_{4j}$
5	$y_{5,1}$	$y_{5,2}$	$\sum_j y_{5j}$
6	$y_{6,1}$	$y_{6,2}$	$\sum_j y_{6j}$
7	$y_{7,1}$	$y_{7,2}$	$\sum_j y_{7j}$
8	$y_{8,1}$	$y_{8,2}$	$\sum_j y_{8j}$
9	$y_{9,1}$	$y_{9,2}$	$\sum_j y_{9j}$
10	$y_{10,1}$	$y_{10,2}$	$\sum_j y_{10j}$
11	$y_{11,1}$	$y_{11,2}$	$\sum_j y_{11j}$
12	$y_{12,1}$	$y_{12,2}$	$\sum_j y_{12j}$
Total C_j	$\sum_i y_{i1}$	$\sum_i y_{i2}$	G

The grand total G is given by:

either summing the row totals R_i over all the 12 rows, i.e.

$$\sum_i \sum_j y_{ij}$$

or   summing the column totals C_j over both columns, i.e.

$$\sum_j \sum_i y_{ij}$$

Example.

Month	i	6	7	8	9	10
Average number of orders taken by one salesman in month i	x_i	9	7	4	5	3
Average order size in month i in £100s	y_i	1	4	2	5	6

Number of salesmen $= b = 4$.

Calculate and give the real meaning of:

(a) $\displaystyle\sum_{i=6}^{10} x_i$

(b) $\displaystyle\sum_{i=6}^{10} bx_i$

(c) $\displaystyle\sum_{i=8}^{10} x_i y_i$

Solution.

(a) $\displaystyle\sum_{i=6}^{10} x_i = x_6 + x_7 + x_8 + x_9 + x_{10}$

$$= 9 + 7 + 4 + 5 + 3$$
$$= 28$$

It is the average total number of orders taken by one salesman between months 6 and 10 inclusive.

(b) $\displaystyle\sum_{i=6}^{10} bx_i = \sum_{i=6}^{10} 4x_i = 4x_6 + 4x_7 + 4x_8 + 4x_9 + 4x_{10}$

$$= 4 \sum_{i=6}^{10} x_i = 4 \cdot 28 = 112$$

It is the total number of orders taken by all four salesmen between months 6 and 10 inclusive.

Note.

$$\sum_i b x_i \text{ may be written as } b \sum_i x_i.$$

(c) $\displaystyle\sum_{i=8}^{10} x_i y_i = x_8 \cdot y_8 + x_9 \cdot y_9 + x_{10} \cdot y_{10}$

$$= 4 \cdot 2 + 5 \cdot 5 + 3 \cdot 6$$
$$= 8 + 25 + 18$$
$$= 51$$

It is the average value of orders (in £100s) taken by one salesman between months 8 and 10 inclusive.

Although additional notations are used in later chapters, they are explained in the appropriate context.

EXERCISES

1. Using the table in the example on page 9 calculate:

(a) $\displaystyle\sum_{i=7}^{9} y_i$

(b) $\displaystyle\sum_{i=6}^{9} y_i^2$

2. Calculate $\displaystyle\sum_{k=8}^{12} k$

3.

i	1	2	3	4	5	6	7	8
x_i	10	4	8	3	4	1	8	5
Z_i	8	2	4	7	5	10	1	3

(a) Is $\displaystyle\sum_{i=4}^{8} x_i < \sum_{i=4}^{8} Z_i$?

(b) Is $\displaystyle\sum_{i=1}^{3} x_i \leqslant 8 + \sum_{i=1}^{3} Z_i$?

(c) Is $11 + \displaystyle\sum_{i=1}^{8} 3x_i \geqslant \sum_{i=1}^{4} x_i Z_i$?

2. *Presentation of Data*

2.1 INTRODUCTION

Numerical data can be communicated in various ways—as the spoken word, in the form of tables, by means of one or two *statistics*, i.e. numbers, which are representative of the data, etc.—and the method of presentation ought to be influenced both by the level of management to which the data is being presented and the use to be made of the information inherent in the data. For example, if the data is to be presented to a market researcher it would be acceptable to transmit it in the form of statistics; if it is to a 'layman' it may be preferable to present the information by graphical means.

A further factor which has an effect on the method of presentation is the volume of the data. If data is limited, then it may be presented in its raw form—just a series of numbers. If there is a large quantity of numerical data, then some other method of presentation is almost certainly required.

This chapter illustrates the two most common ways in which volumes of raw data can be reduced to more compact and comprehensible forms for presentation purposes, namely:

(i) representing the data graphically by means of a *bar chart* or *frequency histogram*;

(ii) calculating statistics which are representative of the data, such as the *mean* and *standard deviation* of a set of observations.

2.2 GRAPHICAL REPRESENTATION

2.2.1 THE BAR CHART (FOR UNGROUPED DATA)

Construction of a bar chart to represent data is usual for data relating to discrete variables and where no grouping of the possible values of the variable occur. This is illustrated using the data in Table 2.1, which refers to the number of private cars sold over a 52 week period by one salesman.

The mass of information in Table 2.1 does not mean very much in its present form except to indicate that the number of cars sold per week varies from zero to eight. One method of making the data more compact and comprehensible is to construct a bar chart as follows:

TABLE 2.1

Car sales per week over 52-week period

```
3  5  2  6  3  5  2  4  8  4
6  4  4  4  5  5  6  1  3  7
2  5  5  7  3  3  5  4  6  6
4  4  4  3  4  0  6  4  3  5
6  5  6  6  4  7  5  4  4  5
5  2
```

Step 1. Identify the range of observed values of the variable. [In this case it is 0–8.]

Step 2. Determine the frequency with which observations (in this case car sales) occur for each of the possible values of the variable. This is done by using 'five-bar gates', which enables raw data to be allocated fairly rapidly to appropriate values of the variable. 'Five-bar gates' are illustrated on Figure 2.1, which has been constructed by allocating each number of Table 2.1 to the appropriate value of the variable of Figure 2.1.

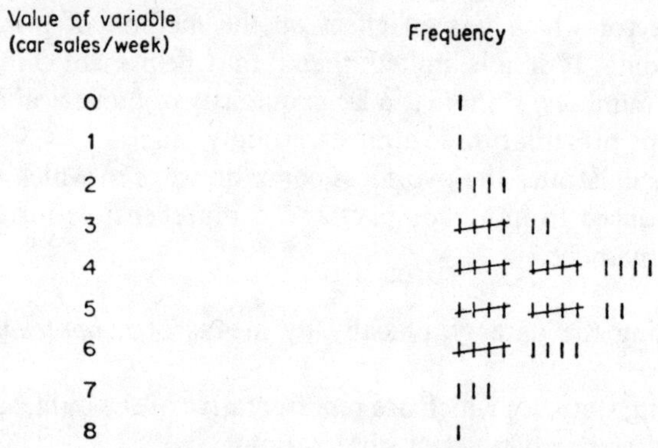

Value of variable (car sales/week)	Frequency
0	I
1	I
2	I I I I
3	++++ I I
4	++++ ++++ I I I I
5	++++ ++++ I I
6	++++ I I I I
7	I I I
8	I

Figure 2.1 'Five-bar gates'

A *frequency distribution* may then be drawn up as in Table 2.2, which indicates the distribution of frequencies of occurrence of the different values of the variable.

The frequency distribution gives a clearer picture of the information inherent in the raw data, first, by structuring the data into an ordered form and, secondly, by showing how the observations are spread or distributed over the range. In this example, it becomes clear that the salesman most often sold either 4 or 5 cars per week and that it was very rare for him to sell either none or many (8). It should be noted that this table could be used as a basis for predicting his future sales, i.e. he is most likely to sell 4 or 5 a week and very unlikely to sell 0 or 8.

TABLE 2.2

Frequency distribution (car sales per week)

Value of variable (No. of cars sold/week)	Frequency
0	1
1	1
2	4
3	7
4	14
5	12
6	9
7	3
8	1
	Total 52

Step 3. Construct the bar chart (Figure 2.2) which is a graphical representation of the frequency distribution.

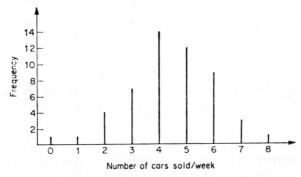

Figure 2.2   Bar chart (car sales/week)

In some texts the tops of the vertical lines of a bar chart are joined to form what is termed a *frequency polygon* but this practice is illogical for discrete variables since a polygon infers that there is a possibility of a frequency occurring for an infinite number of possible values of the variable. In this example only integer values of the variable can occur (the number of cars sold per week). The practice of constructing polygons for ungrouped discrete variables has therefore generally been stopped since it can be misleading.

2.2.2 THE FREQUENCY HISTOGRAM (FOR GROUPED DATA)

(*a*) *Grouped discrete data*. In the previous section ungrouped discrete data was represented graphically by means of a bar chart. When a discrete variable can take a large number of possible values it is usual to group some of these values together and construct a *frequency histogram* to give a graphical representation of the information contained in the data.

The essential feature of a frequency histogram is that a series of rectangles are drawn each on an equal *class width* with the *height* of each rectangle being proportional to the corresponding frequency.

The process to be followed for constructing a histogram for grouped discrete variables is illustrated with reference to Table 2.3 which gives the numbers of cars sold in one month by 100 garages.

TABLE 2.3

Car sales per month

46	52	39	43	69	31	53	52	68	17
6	64	25	88	67	85	57	60	76	60
58	96	67	94	60	73	68	66	41	60
11	38	70	82	40	94	8	86	105	65
79	65	88	54	51	114	59	93	64	31
66	68	37	109	67	59	60	62	41	50
78	97	78	55	74	67	22	40	100	27
20	44	62	72	49	82	54	73	68	38
74	75	57	86	31	82	69	51	53	63
49	70	62	46	69	36	65	83	78	19

Step 1. Identify range. [In this case it is 6–114.]

Step 2. Divide the range into twelve *classes* (or groupings) of width 10 [i.e. 0, 1, 2, . . . 9 are in one group; 10, 11, . . . 19 are in the next group, etc.]. The groupings are usually such that there are between 5 and 20 classes, the actual number depending on the data in the sense that for a given set of data if there are a large number of classes the histogram may be very 'spiky', whereas fewer classes may give a smoother but less detailed representation of the distribution of the data. In this example the classes started at 0, but the first class could equally well have started at any number up to and including the lower limit of the range. There are no hard-and-fast rules for the number of classes or where the first class should start.

Step 3. Use 'five-bar gates' (Figure 2.3) to identify the frequency of observations in each class and then construct the frequency distribution (Table 2.4).

The numbers 0–9, 10–19, 20–29, etc., of Table 2.4 are referred to as *class limits*:

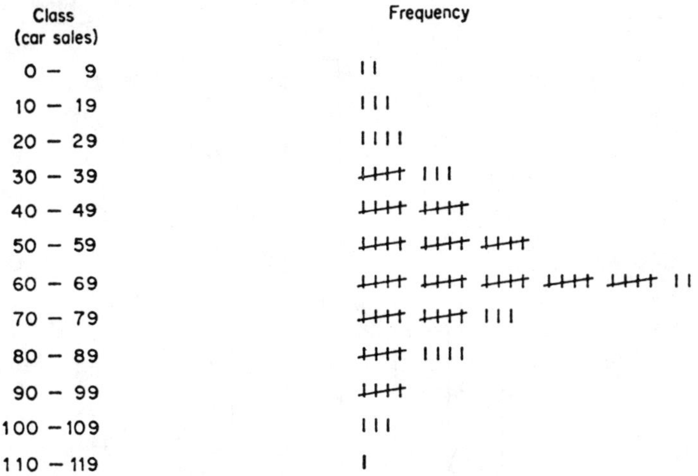

Class (car sales)	Frequency
0 – 9	I I
10 – 19	I I I
20 – 29	I I I I
30 – 39	HHt I I I
40 – 49	HHt HHt
50 – 59	HHt HHt HHt
60 – 69	HHt HHt HHt HHt HHt I I
70 – 79	HHt HHt I I I
80 – 89	HHt I I I I
90 – 99	HHt
100 – 109	I I I
110 – 119	I

Figure 2.3 'Five-bar gates'

0, 10, 20, . . . being the lower class limits; 9, 19, 29, . . . being the *upper class limits*.

Step 4. Construct a frequency histogram (Figure 2.4). The rectangles are drawn with the centre of the base of the rectangle on the *class mark* (*or class mid-point*)

TABLE 2.4

Frequency distribution (car sales per month)
(Grouped discrete data)

Class (Car sales)	Frequency
0–9	2
10–19	3
20–29	4
30–39	8
40–49	10
50–59	15
60–69	27
70–79	13
80–89	9
90–99	5
100–109	3
110–119	1
Total	100

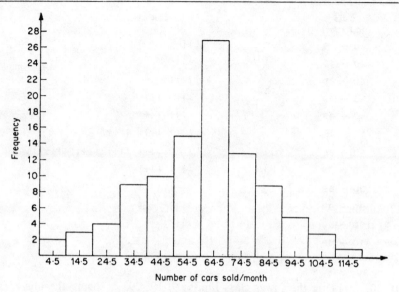

Figure 2.4   Frequency histogram

which is obtained by adding the lower and upper limits of the class and dividing by two. Thus the mid-point for the first rectangle is $(0 + 9)/2 = 4.5$, the second is $(10 + 19)/2 = 14.5$, etc. The base of the rectangle is equal to the class width, which is the difference between the upper and lower class boundaries.

Figure 2.5 is an exploded view of class 10–19 illustrating the class limits, class boundaries, class mark, and class width. The class 10–19 could notionally be thought of as containing all values between 9·5 and 19·5, although it is known that no such decimal values could occur with discrete data. These values are known as the *lower* and *upper class boundaries* respectively. They are convention-

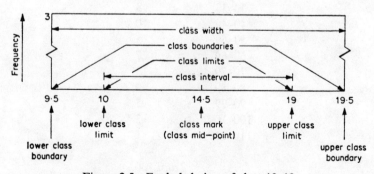

Figure 2.5   Exploded view of class 10–19

ally taken to one more place of decimals than the measurements of data, i.e. one place of decimals in this example. Use of class boundaries allows adjacent rectangles to have common vertical sides.

(*b*) *Grouped continuous data.* Construction of a frequency histogram, which is the usual method of graphically representing continuous variables, will be demonstrated with reference to Table 2.5 which gives the frequency of the distances travelled to the nearest kilometre in one week by 200 salesmen. It will be noted that a rounding-off procedure has to be adopted when measuring continuous variables. In this example the rounding-off is to the nearest kilometre and, in general, the measurement recorded for a continuous variable will only be as accurate as the measuring mechanism. Thus, in this example, a distance recorded as 22 kilometres indicates that the true distance travelled lies in the range $(22-\frac{1}{2})$ to $(22+\frac{1}{2})$ kilometres and any individual distance refers to the occurrence of a value within such a range of values.

Consider the second class of Table 2.5. The numbers 25–49 are the *class limits*,

TABLE 2.5

Frequency distribution (car distance)
(Grouped continuous data)

Class (distance in km)	Frequency
0–24	0
25–49	1
50–74	2
75–99	4
100–124	13
125–149	26
150–174	50
175–199	40
200–224	30
225–249	20
250–274	10
275–299	3
300–324	1
325–349	0
	$\sum f_i = \overline{200}$

25 being the *lower class limit*, and 49 the *upper class limit*. The class 25–49 theoretically includes all values between 24·5 and 49·5 kilometres. The numbers 24·5 and 49·5 are termed *class boundaries* with 24·5 being the *lower class boundary*

and 49·5 the *upper class boundary*. When possible, class boundaries should not coincide with observable values since this creates problems in deciding whether the values lie in the lower or upper class. This can be achieved by defining the boundaries to one more decimal place than the observations.

The rectangles are constructed with the centre of the base of the rectangle on the *class mark* (or *class mid-point*) which is obtained by adding the lower and upper class limits, then dividing this total by two. For example, for the class interval 25–49 the mid-point is (25 + 49)/2 = 37. The base of the rectangle is equal to the class width.

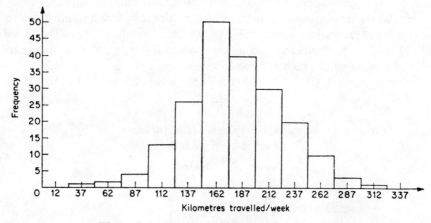

Figure 2.6   Frequency histogram (continuous)

The histogram, Figure 2.6 of the frequency distribution Table 2.5 can now be drawn with the height of the rectangles being equal to the frequencies since the classes are of equal width.

Recapitulating, the steps necessary to construct a frequency bar chart or frequency histogram are:

1. Identify the range of the data.
2. Divide the range into between 5 and 20 groups.
3. Use 'five-bar gates' to determine the frequencies within each group.
4. (*a*) *Bar chart*. Construct the bar chart by drawing vertical lines whose heights are proportional to the frequencies.

(*b*) *Histogram*. Construct the histogram by making the heights of the rectangles proportional to frequencies.

Note. A rectangle on a histogram represents only the frequency of occurrence of events within its class and nothing can be inferred from it as to the distribution of actual events within that class. For discrete data the actual events could only occur on the exact values in the class. For continuous data the actual events could occur at any value in the class.

2.2.3 CUMULATIVE FREQUENCY DISTRIBUTION (OR OGIVE)

A cumulative frequency distribution is a graphical representation of a frequency distribution such that the *cumulative frequency* (sum of the frequencies) of all values less than the upper class boundary of a given class can be obtained directly.

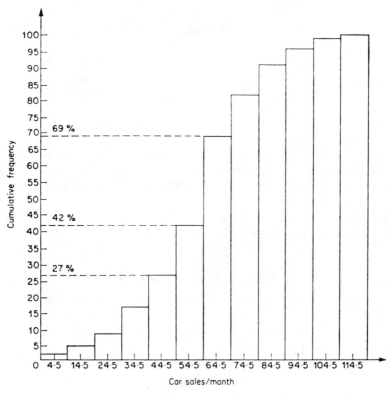

Figure 2.7   Cumulative frequency distribution (discrete variable)

For example, a cumulative frequency distribution for Table 2.3 (car sales/month) is given on Figure 2.7 from which it can be read directly that:

27% of the salesmen sell less than 49·5 cars/month
42%   ,,   ,,   ,,   59·5 cars/month
69%   ,,   ,,   ,,   69·5 cars/month

Before constructing a cumulative frequency distribution it is usual to tabulate cumulative frequencies as in Table 2.6.

TABLE 2.6

Cumulative frequencies (car sales per month)

Class	Frequency	Cumulative frequency
0–9	2	2
10–19	3	5
20–29	4	9
30–39	8	17
40–49	10	27
50–59	15	42
60–69	27	69
70–79	13	82
80–89	9	91
90–99	5	96
100–109	3	99
110–119	1	100

Figure 2.7 is sometimes referred to as a 'less than' ogive since it is cumulative frequencies less than specified classes that are obtained. If the question 'What is the total number of salesmen who sell more than 39·5 cars/month?' is posed then a 'more than' ogive is required.

The appropriate frequencies for a 'more than' ogive are given in Table 2.7 and the 'more than' ogive is given in Figure 2.8.

TABLE 2.7

'More than' ogive—car sales

Class	Frequency	'More than' frequency
0–9	2	100
10–19	3	98
20–29	4	95
30–39	8	91
40–49	10	83
50–59	15	73
60–69	27	58
70–79	13	31
80–89	9	18
90–99	5	9
100–109	3	4
110–119	1	1

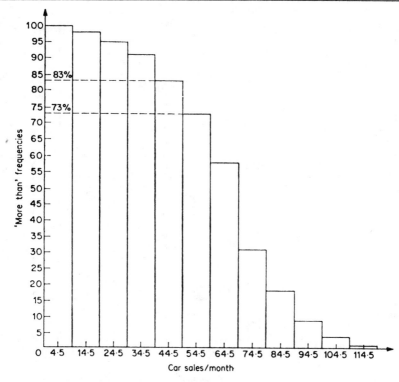

Figure 2.8. 'More than' ogive (car sales)

Note that 'more than' ogives give frequencies greater than the lower class boundary of a given class.

Thus from the 'more than' ogive it can be read directly that:

83% of the salesmen sell more than   39·5 cars/month
73%   „   „   „   49·5 cars/month, etc.

When dealing with grouped continuous data a smooth curve, based on the upper class boundary of the relevant class is sometimes used, rather than a 'stepped' ogive. This is illustrated in Figure 2.9 for the car distance data of Table 2.5. The calculations necessary for construction of Figure 2.9 are given in Table 2.8.

A further point that should be noted is that if the type of ogive or cumulative frequency table is not specified, then it is assumed that a 'less than' ogive or cumulative frequency table is being discussed.

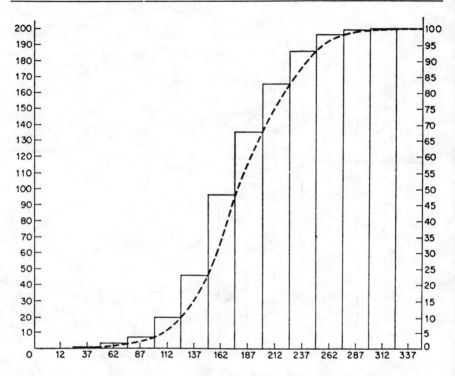

Figure 2.9   Cumulative frequency histogram and cumulative relative frequency (%)

TABLE 2.8

Cumulative frequency table (car distances)

Class	Frequency	Cumulative frequency
0–24	0	0
25–49	1	1
50–74	2	3
75–99	4	7
100–124	13	20
125–149	26	46
150–174	50	96
175–199	40	136
200–224	30	166
225–249	20	186
250–274	10	196
275–299	3	199
300–324	1	200
325–349	0	200

Rather than constructing a cumulative frequency table or distribution, data may be presented with certain advantages in the form of a relative or percentage cumulative frequency table or function.

This is done on Table 2.9 and the cumulative frequency function is given on Figure 2.9 using the scale on the right-hand side.

TABLE 2.9

Relative cumulative frequency table (car distances)

Class	Cumulative relative frequency (%)
0–24	$\frac{0}{200} \times 100 = 0\cdot0$
25–49	$\frac{1}{200} \times 100 = 0\cdot5$
50–74	$\frac{3}{200} \times 100 = 1\cdot5$
75–99	$\frac{7}{200} \times 100 = 3\cdot5$
100–124	$\frac{20}{200} \times 100 = 10\cdot0$
125–149	$\frac{46}{200} \times 100 = 23\cdot0$
150–174	$\frac{96}{200} \times 100 = 48\cdot0$
175–199	$\frac{136}{200} \times 100 = 68\cdot0$
200–224	$\frac{166}{200} \times 100 = 83\cdot0$
225–249	$\frac{186}{200} \times 100 = 93\cdot0$
250–274	$\frac{196}{200} \times 100 = 98\cdot0$
275–299	$\frac{199}{200} \times 100 \doteq 99\cdot5$
300–324	$\frac{200}{200} \times 100 = 100\cdot0$
325–349	$\frac{200}{200} \times 100 = 100\cdot0$

The advantage of the percentage cumulative frequency function is that 'percentages' can be obtained directly from the function. Thus it is seen from Figure 2.9 that:

48% of the time a salesman travels less than 174·5 kilometres/week
90%   ,,   ,,   ,,   ,,   240   ,,

It will be noticed from the above that when dealing with a continuous variable, it is possible to interpolate directly from the percentage ogive.

2.2.4 OTHER METHODS OF GRAPHICAL REPRESENTATION

Some further methods for presentation of data used especially for publicity purposes include:

(*a*) a bar chart where the rectangles are separated (Figure 2.10);
(*b*) a component bar chart (Figure 2.11);
(*c*) a pie chart (Figure 2.12);
(*d*) a pictograph (Figures 2.13 and 2.14); and
(*e*) a Z chart (Figure 2.15).

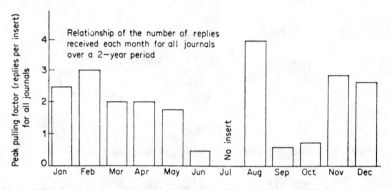

Figure 2.10   Bar chart (source: *The Financial Times*, 25.2.69)

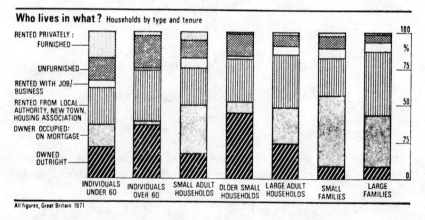

Figure 2.11   Component bar chart (source: *The Economist*)

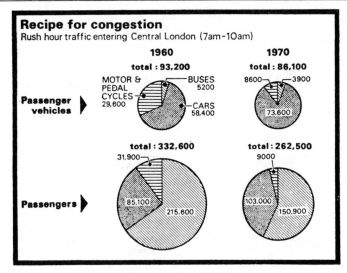

Figure 2.12   Pie chart (source: *The Economist*)

The reader will doubtless be familiar with such presentations. A few points can be made however about construction of such figures.

The component bar chart is constructed such that the areas of the rectangles within the total rectangle are proportional to the quantities being compared. For the pie chart, slices of the pie correspond in size with the various quantities being compared. The use of a pictograph may be misleading. If the units of the pictograph are all the same size (as in Figure 2.13) comparisons may not be too

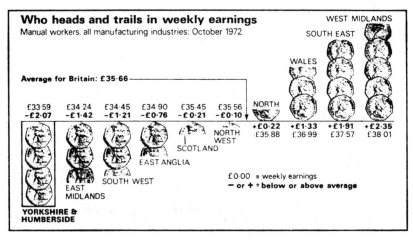

Figure 2.13   Pictograph (source: *The Economist*)

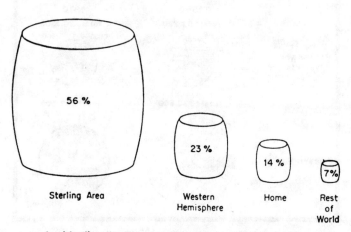

A misleading diagram, proportions being based on *heights*, but the
reader is more likely to base his impressions on *volumes*

Figure 2.14   Pictograph (source: *Facts from Figures*, Pelican Original 1951.
Copyright © M. J. Moroney 1951)

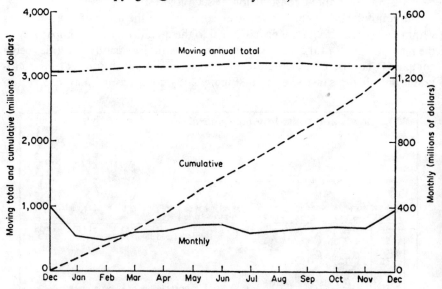

Sales of Sears Roebuck and Company; Monthly, Cumulative, and
Moving Annual Total, 1953. Data from *Survey of Current Business*, February
1953, p. S–10, and February 1954, p. S–10.

Figure 2.15   Z chart (source: Croxton, F. E. and Cowden, D. J. *Applied General
Statistics*, 2nd edition. Prentice-Hall/Pitman, 1960)

confusing. Serious difficulties can result if comparisons are made of the size of the units since there may be uncertainty as to the particular size to be compared. For example, given Figure 2.14, if it is the heights that are being compared then they are in the ratio 2:1, if it is area the ratio is 4:1, and if it is volume the ratio is 8:1. Great care must be taken when examining such pictographs.

A Z chart consists of three curves on the same axis as illustrated on Figure 2.15. For the example illustrated on the figure these are:

(*a*) monthly sales figures;

(*b*) cumulative sales figures;

(*c*) moving annual total, i.e., it is the total annual sales for the twelve month period ending in a given month.

The Z chart is constructed with two vertical scales; the left-hand one being used for plotting cumulative sales and moving annual total; while the right-hand scale is used for plotting monthly sales. The necessity for two scales is due to the differences in the magnitude of numbers used for monthly sales compared to cumulative and moving annual figures.

The above methods of graphical representation in general are not sufficiently accurate or convenient for statistical work and so, in practice bar charts or histograms are used.

2.3 REPRESENTATIVE STATISTICS

Graphical representations of data, such as the forms illustrated in Section 2.2 often give a very clear picture of the information inherent in data. However, it may be more useful to refer to data by one or two numbers and it is then possible to communicate verbally rather than on paper as is necessary with graphical representations.

The numbers (statistics) used to represent data are generally those which measure *average* (*central tendency*) and those which measure *dispersion* (spread of a set of numbers), although for special purposes other statistics may also be calculated.

The average of a set of numbers indicates a value around which the data may be considered to cluster—hence the term central tendency—and may be taken to be typical of the data. The dispersion of data gives an indication of the spread of the data and the range (*see* Section 2.2) is one measure of this dispersion. Other measures of dispersion will be introduced in Section 2.3.2.

2.3.1. MEASURES OF AVERAGE

There are three main measures of average—mean, median, and mode. Which one should be used? Before this question can be answered it is necessary to stipulate certain properties which an ideal measure of average should possess. Intuitively the following properties seem reasonable:

1. It should be easy to calculate (for practical purposes).
2. It should be capable of objective definition (i.e. its value should not depend on any subjective decisions by the individual calculating it).
3. It should make use of all members of the set of numbers (otherwise 'information' contained in the set is lost).
4. It should be usable mathematically in other statistical calculations. For example, it should be possible to combine the averages obtained from a number of groups of data to give the overall average which would have been obtained had the groups been first combined, and then the average of the combined groups calculated.

Each measure is now considered.

(a) *Mean.* This is the most commonly used measure of average and it is sometimes referred to as the *arithmetic mean* (or arithmetic average). The mean is calculated by adding the values in the data and then dividing by the *number* of values in the data.

Given the following data on the monthly expenses of 10 salesmen:

£100, £125, £150, £125, £200, £150, £175, £250, £150, £150

then the mean monthly expense is:

$$\frac{100 + 125 + 150 + 125 + 200 + 150 + 175 + 250 + 150 + 150}{10}$$

$$= \frac{1,575}{10}$$

$$= £157 \cdot 50$$

Expressing this mean in algebraic notation; if there are n observations x_1, x_2, x_3, ... x_n, then the mean of these observations, denoted by m is:

$$m = \frac{x_1 + x_2 + x_3 + \ldots + x_n}{n}$$

$$\therefore m = \frac{\sum_{i=1}^{n} x_i}{n} \qquad [2.1]$$

In some texts $\bar{x}$ is used to denote the arithmetic mean. Here we usually use m, but employ $\bar{x}$ and $\bar{y}$ where the means of two variables are required.

If the salesmen's expenses data was presented as a frequency distribution, such as Table 2.10, then the mean of the data is:

TABLE 2.10

Frequency distribution (salesmen's expenses)

Salary	Frequency
100	1
125	2
150	4
175	1
200	1
250	1

$$\sum f_i = 10$$

$$\frac{1 \times 100 + 2 \times 125 + 4 \times 150 + 1 \times 175 + 1 \times 200 + 1 \times 250}{10}$$

$$= £157{\cdot}50$$

In notation form when there are k classes (or groupings) the equation is:

$$m = \frac{f_1 x_1 + f_2 x_2 + \ldots + f_k x_k}{f_1 + f_2 + f_3 + \ldots + f_k}$$

$$= \frac{\sum\limits_{i=1}^{k} f_i x_i}{\sum\limits_{i=1}^{k} f_i}$$

$$\therefore m = \frac{\sum\limits_{i=1}^{k} f_i x_i}{n} \qquad \qquad [2.2]$$

where: f_i = frequency of the ith class

$\quad\quad\quad x_i$ = value of the ith class

$\quad\quad\quad n$ = total number of observations $\left(= \sum\limits_{i=1}^{k} f_i \right)$

If the data is in the form of a grouped frequency table, then the above formula for the mean is still appropriate if x_i is taken as the value of the class mark of the ith class. Note, however, that the mean calculated from grouped data using Equation 2.2 may not be identical with the mean of the data in ungrouped form. This is due to the effect of grouping, i.e., loss of detail, thus assuming that all observations in a class have the value of the class mark.

The arithmetic mean of the grouped frequency table for car sales (Table 2.4) is calculated in Table 2.11 and gives:

$$m = \frac{6,060}{100} = 60{\cdot}6 \; cars \; sold \; per \; month$$

TABLE 2.11

Calculation of mean (car sales per month)

Class (Car sales)	Frequency, f_i	Mid-point, x_i	$f_i x_i$
0–9	2	4·5	9·0
10–19	3	14·5	43·5
20–29	4	24·5	98·0
30–39	8	34·5	276·0
40–49	10	44·5	445·0
50–59	15	54·5	817·5
60–69	27	64·5	1,741·5
70–79	13	74·5	968·5
80–89	9	84·5	760·5
90–99	5	94·5	472·5
100–109	3	104·5	313·5
110–119	1	114·5	114·5
	$\sum f_i = 100$		$\sum f_i x_i = 6{,}060·0$

$$m = \frac{\sum f_i x_i}{\sum f_i} = \frac{6{,}060}{100} = 60·6$$

At the start of this Section, four desirable properties for a measure of average were identified. The mean meets all of these properties for ungrouped data, but its objectivity can be compromised with grouped frequency distributions (see comments later about the median). Perhaps the most difficult property to accept at first sight is its property of mathematical manipulation. For example, two means can be combined to give an overall mean using the formula:

$$M = \frac{m_1 n_1 + m_2 n_2}{n_1 + n_2}$$

where: m_1 is the mean of n_1 observations,
m_2 is the mean of n_2 observations.

This is now illustrated as follows. An advertising campaign was run for a product in three journals, A, B, and C in month 1 and in journals A and B in month 2. The response data in terms of the number of enquiries received is given in Table 2.12. For the first month the average number of responses per journal was $(75 + 50 + 60)/3 = 61\frac{2}{3}$ and for the second month it was $(50 + 50)/2 = 50$.

The overall mean is then $[61\frac{2}{3}(3) + 50(2)]/(3 + 2) = 57$ and this would be the same had the data first been combined and then the mean calculated:

mean = $(75 + 50 + 60 + 50 + 50)/5 = 57$ enquiries.

The mean is usually used as the standard for calculating measures of dispersion and, in fact, the bulk of statistical theory is based upon the mean.

TABLE 2.12

Response rate

	Enquiries *Month 1*	*Enquiries* *Month 2*
Journal A	75	50
Journal B	50	50
Journal C	60	

Perhaps the main disadvantages of the mean as a measure of average are that:

(*a*) it can result in a value which cannot be obtained in practice (e.g. it is impossible to sell 60·6 cars/month) and hence it is not strictly typical of the data; and

(*b*) it can be considerably influenced by extreme values in small samples of data.

For example, given the following set of numbers:

4, 5, 5, 6, 7, 7, 8

the average of them is 6. If, however, the upper value of 8 changed to 29, then the average would be 9, which is not a 'typical' value of the set of numbers 4, 5, 5, 6, 7, 7, 29.

(*b*) *Median.* The median of a set of observations is the central value when the observations are ranked in order of magnitude. If there are an even number of observations, then the median is taken as the mean of the central pair.

The following are salesmen's expenses ranked in order of magnitude:

£100, £125, £125, £150, £150, £150, £150, £175, £200, £250

and the median is $(150 + 150)/2 = £150$

The median for the following set of numbers:

1, 3, 7, 9, 9, 10, 11, 11, 12, 15, 20

is 10. Here there are five numbers below 10 and five above.

When dealing with a grouped frequency distribution the median may be obtained either by interpolation within the median class or, more conveniently and practically, by accepting the median class mark as the median.

For example, from Table 2.8 the median lies within the class 175–199 since it contains the 100th and 101st values of the set of 200 values. The median may therefore be taken as 187, the class mark of that class.

The median does not meet all the desirable properties of an average. The following points can be made:

1. It is easy to calculate.

2. Its objectivity can be compromised with grouped frequency distributions due to the necessary subjective choice of groupings which can affect the value obtained.

3. It is a function of the *number* of values in the set although it does not make use of the actual values.

4. It is not capable of much useful mathematical manipulation. For example, it is not possible to calculate the joint median of two separate sets of data from the two independent medians. It would be necessary to combine the data and then order it so as to find the middle value.

If the grouping had been as Table 2.13, which is the raw data of Table 2.3

TABLE 2.13

Frequency distribution (car sales)

Class (*Sales*)	Frequency
5–14	3
15–24	4
25–34	5
35–44	11
45–54	13
55–64	17
65–74	23
75–84	10
85–94	8
95–104	3
105–114	3
Total	100

grouped into class intervals different to those in Table 2.4, then the median would be 59·5.

The point to be made is that the median like the mean is only objectively defined for grouped data once the grouping has been made.

(c) *Mode.* The mode is the most frequently occurring value of the variable in a set of observations, i.e., that value occurring at the peak of a histogram.

Referring again to the salesmen's expenses data:

£100, £125, £125, £150, £150, £150, £150, £175, £200, £250

the mode is £150.

Given the following numbers:

1, 2, 3, 5, 7, 9, 9, 9, 9, 10, 11, 11, 11, 11, 12, 15, 20, 23, 29

there are *two* modes, 9 and 11, and this set of data is termed *bi*modal.

When there are *several* modes not necessarily of the same frequency the data is said to be *multi*modal.

When dealing with grouped data in the form of a frequency distribution the mode is usually identified in terms of the modal class or as the class mark of the modal class although interpolation is sometimes done.

Thus, for Table 2.5 it is seen that the mode is represented by the class 150–174, or by the class mark 162.

As was the case for calculation of the median from a grouped frequency distribution the mode can be changed by alteration of the class intervals and limits.

The properties of the mode are:

1. It is easy to calculate.
2. It is objectively defined only for ungrouped data.
3. It does not make use of all members of the group.
4. It is not capable of mathematical manipulation. For example a joint mode cannot be found for two sets of data without first combining the data (c.f. median).

It is easy to demonstrate that the mode does not represent all members of a set of observations, since it is clear that provided the frequency of the original modal class remains the largest then all other frequencies can be altered without changing the modal class.

The possibility of multimodes lessens the attraction of the mode as a measure of average since both the mean and median produce unique numbers for any set of observations.

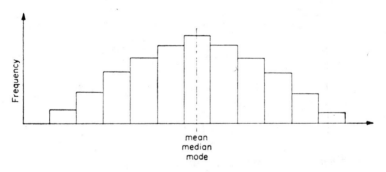

Figure 2.16   Frequency histogram (symmetrical)

(*d*) *Graphical representation of mean, median, and mode.* The frequency histogram of Figure 2.16 is symmetrical and when this is the case for a frequency distribution the mean, median, and mode all take the same value.

On the other hand, the frequency histogram of Figure 2.17 is termed a skewed distribution and in this case the mean, median, and mode always take different values. Figure 2.17 is in fact an illustration of positive skewness since the tail of the histogram lies to the right of the mode. If the tail was to the left of the mode this would be termed negative skewness.

When dealing with skewed distributions the median always lies between the mean and the mode.

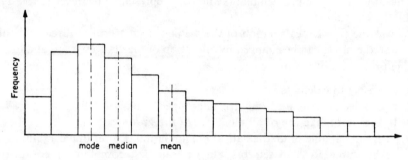

Figure 2.17   Frequency histogram (skewed)

2.3.2 MEASURES OF VARIABILITY

Although it has now been demonstrated how averages can be determined no single average by itself indicates the variability (or dispersion) of the data to which the average refers. For example, Figure 2.18 illustrates two histograms for both of which the mean, median, and mode are the same: however it is clear that

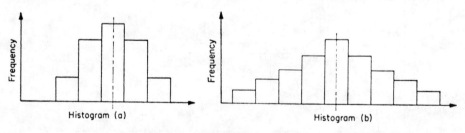

Figure 2.18

there is more spread in histogram (*b*) than in histogram (*a*) where the data is closely clustered around the central value.

Presentation of data in the form of a histogram clearly demonstrates the variability inherent in the data. In this section the three most common calculated statistics used to indicate the variability of data will be discussed in some depth; range, variance, and standard deviation. Less commonly used measures, namely the coefficient of variation, and quantiles are outlined in Part (*d*) of this section.

The criteria applied to measures of average, namely (1) ease of calculation, (2) objectivity, (3) representation of all members of the set of values, and (4) ease of mathematical manipulation, can equally well be applied to measures of variability.

(*a*) *Range*. This measure of variability has been introduced in Section 2.2 with reference to the construction of a bar chart or histogram. The range is simply the difference between the largest and smallest values in the set of data.

For example, for the salesmen's expenses referred to in Section 2.3.1 the range is:

£250–£100 = £150

When dealing with grouped data there are two procedures which are adopted for determining the range:

1. Range = class mark of highest class – class mark of lowest class.
2. Range = upper class boundary of highest class – lower class boundary of lowest class.

Applying these two procedures to the cars sold per salesmen frequency table (Table 2.4) gives:

Method 1. Range = 114·5 – 4·5 = 110
Method 2. Range = 119 – 0 = 119

Using the data in Table 2.5 for car distance the range is either 312 – 37 = 275 kilometres (taking the class mark) or 324·5 – 24·5 = 300·0 kilometres (taking class boundaries).

Referring to the criteria for a 'good' measure of variability it can be said that the range is:

1. Easy to calculate.
2. Objectively defined only for ungrouped data.
3. Makes use of only two members of the set although it does investigate all the values of the set.
4. Only capable of limited mathematical manipulation. (For example, for small sample sizes an estimate of the variance (*see* (*b*)) may be obtained from the range.)

The main disadvantage of the range as a measure of variability is associated with the fact that it is making use of only two members of the set of data, since this can (*a*) result in 'freak' values giving a misleading impression of variability and (*b*) result in the same range for two sets of data when there is clearly much greater variability in one set than in another. These two points can be demonstrated as follows.

Suppose the returns from six different sales areas gave the following numbers of orders obtained in a three month period:

15, 22, 30, 36, 40, 48, 57, 60, 71, 102

The range of this set would be 102 – 15 = 87. If on only a small proportion of occasions it is known that sales in excess of say 100 are obtained, then omission of the return of 102 would give a much more 'realistic' picture of the spread of return. In the above case the range (omitting 102) is $71 - 15 = 56$. Of course one cannot, in general, be so sure of the 'status' of extreme values as in the above example.

Figure 2.19 illustrates two histograms where the ranges are equal, but it is clear that there is greater variability in (a), where the frequencies are more or less equally spread throughout the range, than in (b), where the frequencies are clustered around a central value with only a few at the ends of the range.

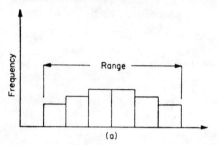 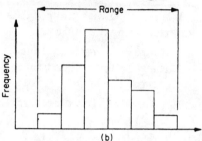

Figure 2.19

It should be noted that the calculation of range does not depend in any way on a measure of average. Thus it cannot be implied, for instance, that an average lies at the mid-point of the range. In fact the range gives little information about how the observations in general are dispersed about an average.

(b) *Variance*. A measure of the spread of numerical data about its mean is given by the variance, which is the mean of the sum of the squared deviations of individual values from the mean of the set of values. The formulae for the variance, denoted by s^2, are:

$$s^2 = \frac{\sum_{i=1}^{n} (x_i - m)^2}{n} \quad \text{(for ungrouped data)} \qquad [2.3]$$

and

$$s^2 = \frac{\sum_{i=1}^{k} f_i(x_i - m)^2}{\sum_{i=1}^{k} f_i} \quad \text{(for grouped data)} \qquad [2.4]$$

where: f_i = frequency of the ith class
$\quad\quad x_i$ = value of the ith class
$\quad\quad m$ = mean of the set of values
$\quad\quad n$ = total number of observations
$\quad\quad k$ = number of classes for grouped data

[In certain situations a modification is made to Equations [2.3] and [2.4] (*see* page 43).]

Calculation of the variance of the salesmen's expenses data could then proceed as follows:

Step 1. Calculate mean of data, i.e.

m = £157·50 (from Section 2.3.1 (*a*))

Step 2. From each value subtract the mean, and square this difference, then add together the squared values.

$(100 - 157·5)^2 + (125 - 157·5)^2 + (125 - 157·5)^2 + (150 - 157·5)^2 +$
$(150 - 157·5)^2 + (150 - 157·5)^2 + (150 - 157·5)^2 + (175 - 157·5)^2 +$
$(200 - 157·5)^2 + (250 - 157·5)^2 = 16,312·5$

Step 3. Divide the answer obtained in Step 2 by the total number of values, i.e. the total frequency. This gives the variance:

$$s^2 = \frac{16,312·5}{10}$$
$$= 1,631·25$$

The calculations of these three steps can be made compact by using a tabular form, Table 2.14, where the calculations performed above are repeated.

TABLE 2.14

Calculation of variance (salesmen's expenses)

Salary x_i	$(x_i - m) = (x_i - 157·5)$	$(x_i - m)^2 = (x_i - 157·5)^2$
100	−57·5	3,306·25
125	−32·5	1,056·25
125	−32·5	1,056·25
150	−7·5	56·25
150	−7·5	56·25
150	−7·5	56·25
150	−7·5	56·25
175	17·5	306·25
200	42·5	1,806·25
250	92·5	8,556·25
	Total	16,312·5

$$s^2 = \frac{\Sigma (x_i - m)^2}{n} = \frac{16,312·5}{10} = 1,631·25$$

The units of variance, for the salesmen's expenses example, are £^{2} (pounds squared). This creates difficulties in interpreting the variance but these may be reduced by taking the square root of the variance. The statistic so formed is termed the *standard deviation*. The standard deviation now gives a measure of data spread in the same units as the mean and for this example the standard deviation, denoted by s, is

$$s = \sqrt{1,631 \cdot 25} \qquad\qquad = £40 \cdot 39$$

The following information has now been calculated for the salesmen's expenses data:

mean, $m = £157 \cdot 5$ standard deviation, $s = £40 \cdot 39$

These two statistics give a measure of average expenses and a measure of the spread of expenses.

Thus, if for another set of salesmen mean expenses were £157·5 with an associated standard deviation of £45 it is clear that there is more variability in the second set than in the first.

This is illustrated on Figure 2.20.

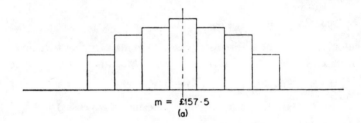

m = £157·5
(a)

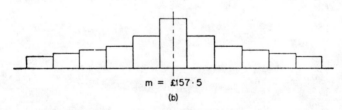

m = £157·5
(b)

Figure 2.20

The standard deviation (s.d.) will be covered in more detail in the next section but the reader may wish to refer forward at this stage to Figure 3.4 of Section 3.4.2 where it is illustrated that for a particular symmetrical probability distribution (the normal distribution):

about 68% of all observations can be expected to lie within ± 1 s.d. of the mean about 95% of all observations can be expected to lie within $\pm$ 2 s.d. of the mean

about 99% of all observations can be expected to lie within $\pm$ 3 s.d. of the mean

The uses of mean and variance, or standard deviation, will become apparent in the later chapters, especially Chapter 4 which deals with sampling.

Calculation of the variance of the cars sold per month of Section 2.2.2 (Table 2.4) is given on Table 2.15. Note that when dealing with grouped data the x_i values are taken as the class marks (mid-points of class interval).

TABLE 2.15

Variance calculation (car sales per month)

Class (Car sales)	Class mark x_i	$(x_i - m) = (x_i - 60{\cdot}6)$	$(x_i - m)^2 = (x_i - 60{\cdot}6)^2$	f_i	$f_i(x_i - m)^2$
0–9	4·5	−56·1	3,147·21	2	6,294·42
10–19	14·5	−46·1	2,125·21	3	6,375·63
20–29	24·5	−36·1	1,303·21	4	5,212·84
30–39	34·5	−26·1	681·21	8	5,449·68
40–49	44·5	−16·1	259·21	10	2,592·10
50–59	54·5	−6·1	37·21	15	558·15
60–69	64·5	3·9	15·21	27	410·67
70–79	74·5	13·9	193·21	13	2,511·73
80–89	84·5	23·9	571·21	9	5,140·89
90–99	94·5	33·9	1,149·21	5	5,746·05
100–109	104·5	43·9	1,927·21	3	5,781·63
110–119	114·5	53·9	2,905·21	1	2,905·21
			Totals	100	48,979·00

$$s^2 = \frac{\Sigma f_i(x_i - m)^2}{n} = \frac{48{,}979{\cdot}00}{100}$$
$$= 489{\cdot}79$$

Table 2.15 illustrates that when dealing with large sets of numbers the variance calculation can become tedious, especially having to square 'odd' numbers which are quite large. For example it is easy to square 3 (i.e. $3 \times 3 = 9$) but when it comes to squaring 26·1 (i.e. $26{\cdot}1 \times 26{\cdot}1 = 681{\cdot}21$) as for class 30–39 of Table 2.15 the arithmetic becomes tiresome. There are short-cut formulae completely equivalent to Formulae 2.3 and 2.4 which can be used and which may reduce the possible difficulties of squaring $(x_i - m)$ values. These are:

$$s^2 = \frac{\sum_{i=1}^{n} x_i^2}{n} - m^2 \text{ (for ungrouped data)} \qquad [2.5]$$

$$s^2 = \frac{\sum\limits_{i=1}^{k} f_i x_i^2}{\sum\limits_{i=1}^{k} f_i} - m^2 \text{ (for grouped data)} \qquad [2.6]$$

Formula 2.6 will be demonstrated with reference to the data of Table 2.4 and the result can be compared with the calculations of Table 2.15. The relevant calculations are given in Table 2.16. In this example 'difficult' numbers still have

TABLE 2.16

Variance calculation (car sales per month)

Class (Car sales)	Class mark, x_i	x_i^2	f_i	$f_i x_i^2$
0–9	4·5	20·25	2	40·50
10–19	14·5	210·25	3	630·75
20–29	24·5	600·25	4	2,401·00
30–39	34·5	1,190·25	8	9,522·00
40–49	44·5	1,980·25	10	19,802·50
50–59	54·5	2,970·25	15	44,553·75
60–69	64·5	4,160·25	27	112,326·75
70–79	74·5	5,550·25	13	72,153·25
80–89	84·5	7,140·25	9	64,262·25
90–99	94·5	8,930·25	5	44,651·25
100–109	104·5	10,920·25	3	32,760·75
110–119	114·5	13,110·25	1	13,110·25
		Totals	100	416,215·00

$$s^2 = \frac{\Sigma f_i x_i^2}{n} - m^2$$

$$= \frac{416,215\cdot00}{100} - (60\cdot6)^2$$

$$= 489\cdot79 \qquad \text{where } m = 60\cdot6 \text{ is obtained from Section 2.3.1}(a).$$

to be squared due to the 'awkwardness' of the class marks. If the class marks had been integers, then computational difficulties would have been reduced.

Although Formula [2.6] was illustrated by means of its application to grouped discrete data (car sales) it can be used for grouped continuous data.

In general when a variance calculation is required it is quicker to apply Formula [2.5] or [2.6] than the original Formulae [2.3] and [2.4]. The equivalence of these formulae is shown in Appendix A.

With reference to the desirable properties of a measure of variability the properties of the variance are:

1. It is not very easily calculated, or at least it is less easy to calculate than the range.
2. It is objectively defined except of course for grouped data.
3. It makes use of all the values in the set.
4. It is useful for mathematical manipulation. (For example it is possible to add variances of independent variables.)

The main disadvantage of the variance as a measure of variability is that the units of variance are difficult to interpret. This, however, is avoided by using the standard deviation which has the same units as the mean. In practice it is found that the algebraic manipulations which can be performed with the variances of sets of data are so useful that the variance (or the related standard deviation) is the most generally used measurement of variability.

(*c*) *Standard deviation.* This statistic was introduced when discussing the variance measure of dispersion. The standard deviation, which is the square root of the variance, is a number which gives a measure of the spread of data about its mean in the same units as the mean.

For the variance calculations of Tables 2.14 and 2.15 the respective standard deviations are:

(i) Standard deviation, $s = \sqrt{\text{variance } (s^2)}$

$$= \sqrt{\frac{\Sigma(x_i - m)^2}{n}}$$

$$= \sqrt{1,631 \cdot 25} = £40 \cdot 39$$

(ii) Standard deviation, $s = \sqrt{\frac{\Sigma f_i(x_i - m)^2}{n}}$

$$= \sqrt{489 \cdot 79}$$

$$= 22 \cdot 13 \text{ cars}$$

Note. The summation signs have no limits specified in the above formula. As mentioned in Section 1.2.6 this is the usual convention when the limits are clear.

The properties of the standard deviation are:

1. It is not easily calculated relative to the range.
2. It is objectively defined except for grouped data.
3. It is calculated from all the values in the set.
4. It is not capable of algebraic manipulation.

(although its square, the variance, is capable of algebraic manipulation).

(*d*) *Coefficient of variation; quantiles.* The most commonly used measure of the variability of a statistical distribution is its standard deviation. Two further statistics which are occasionally calculated are the coefficient of variation and quantiles.

(i) *Coefficient of variation.* In Section 2.3.2(*c*) the standard deviation of the number of cars sold/week by a salesman was calculated to be 22·13, which is a measure of the variability of sales from week to week. This variability really only makes sense if the mean number of sales/week is also known (60·6 in the example) since the magnitude of the variability can then be considered in the light of the mean sales. The coefficient of variation gives such a measure of variability and is defined as:

$$\text{Coefficient of variation, } v = \frac{\text{standard deviation}}{\text{mean}} = \frac{s}{m} \qquad [\nu = \text{Greek nu}]$$

which is usually expressed as a percentage, e.g.

the coefficient of variation of cars sold/week is $\dfrac{22 \cdot 13}{60 \cdot 6} \times 100 = 36 \cdot 5\%$

(ii) *Quantiles.* Quantiles may be determined by using a cumulative relative frequency histogram (*see* Figure 2.9). Quantiles are divided into *quartiles*, *deciles*, and *percentiles*. As described earlier in the chapter the median value of a variable divides the frequency distribution into two equal parts. Quartile values of the variable divide the distribution into four equal parts, the respective quartiles being termed the first, second, and third quartiles. The values of the variable which divide the distribution into ten equal parts are termed deciles, while those which divide it into one hundred equal parts are termed percentiles.

2.3.3. THE USE OF A PROVISIONAL MEAN FOR CALCULATING THE MEAN AND VARIANCE

Although Formulae [2.1] and [2.2] give directly the mean of a set of values it is clear from the calculation of Table 2.11 that 'large' numbers may be involved in calculating $f_i x_i$ and a 'large' number is likely to occur for the summation $\Sigma f_i x_i$. It would be preferable if 'small' numbers could be used since this would speed calculations and reduce the possibility of errors. The use of a provisional mean does just this.

Formulae [2.5] and [2.6] may considerably reduce the awkwardness of the variance calculation compared to Formulae [2.3] and [2.4]. However, from Table 2.16 it is seen that 'large' numbers may still be involved in calculation of the variance. Again, it would be helpful if 'small' numbers were used and again this is achieved by use of a provisional mean.

TABLE 2.17

Mean and variance calculation—use of provisional mean for (car sales per month)

Class (Car sales)	Class mark, x_i	$u_i = \dfrac{x_i - 64 \cdot 5}{10}$	u_i^2	f_i	$f_i u_i$	$f_i u_i^2$
0–9	4·5	−6	36	2	−12	72
10–19	14·5	−5	25	3	−15	75
20–29	24·5	−4	16	4	−16	64
30–39	34·5	−3	9	8	−24	72
40–49	44·5	−2	4	10	−20	40
50–59	54·5	−1	1	15	−15	15
60–69	64·5	0	0	27	−102	0
70–79	74·5	1	1	13	13	13
80–89	84·5	2	4	9	18	36
90–99	94·5	3	9	5	15	45
100–109	104·5	4	16	3	12	48
110–119	114·5	5	25	1	5	25
			Total	100	63	505
					−102	
			Total		−39	

The application of a provisional mean for calculation of the mean and variance of data proceeds as follows:

The formulae for calculating the mean and variance of a set of numbers using a provisional mean, denoted by a, are:

$$m = c\bar{u} + a \quad \text{(for ungrouped or grouped discrete or continuous data)} \qquad [2.7]$$

$$s^2 = c^2 \left[\frac{\Sigma u_i^2}{n} - (\bar{u})^2 \right] \text{(for ungrouped data)} \qquad [2.8]$$

$$s^2 = c^2 \left[\frac{\Sigma f_i u_i^2}{n} - (\bar{u})^2 \right] \text{(for grouped discrete or continuous data)} \qquad [2.9]$$

where: m, s^2, f_i, and n are as previously defined

c = class width

u_i = a new variable which is derived from the x_i value and is

$$u_i = \frac{x_i - \text{provisional mean, } a}{c}$$

$\bar{u}$ = mean of the new variable, $\bar{u} = \Sigma u_i / n$ (for ungrouped data)

= $\Sigma f_i u_i / n$ (for grouped data)

Formulae [2.7]–[2.9] are completely equivalent to the basic formulae (Formulae [2.1]–[2.6]) and their equivalence is shown in Appendix A.

The provisional mean is in effect a guess at what the actual mean of the data will be and it is obtained by inspection of the raw data, preferably in the form of a frequency distribution. Of course, different people may choose a different provisional mean but this in no way affects the end result.

As an example consider the frequency distribution for car sales (Table 2.4). By inspection of this table one might guess that the mean number of cars sold is between 60 and 69, say 64·5, the class mark. Therefore let $a = 64·5$ and $c = 10$. Once a provisional mean has been identified, Table 2.17 can be constructed and the following calculations performed:

The mean of the new variable, u is:

$$\frac{\Sigma f_i u_i}{n} = \frac{-39}{100} = -0·39$$

and the mean of the original variable (the xs) is, from Formula [2.7]:

$$m = c\bar{u} + \text{provisional mean}$$
$$= 10 (-0·39) + 64·5$$
$$= 60·6$$

From Formula [2.9]:

$$\text{The variance (of the } x\text{s)} = c^2 \left[\frac{\Sigma f_i u_i^2}{n} - \bar{u}^2 \right]$$
$$= 100 \left[\frac{505}{100} - (-0·39)^2 \right]$$
$$= 489·79$$

It should be noted that in the sixth column of Table 2.17 the upper half of the column has negative numbers while the lower numbers are positive. The value of $f_i u_i$ for the class in which the provisional mean lies will be zero, and it is usual to use this 'blank space' in the $f_i u_i$ column to put the summation of the 'top half' of the column and then subtract this from the summation of the 'lower half' at the foot of the $f_i u_i$ column, as is done in Table 2.17.

The reader would be strongly recommended to use a provisional mean when calculating the mean and variance of a set of data. Even although it may seem, at first sight, a long process it will almost certainly result in a speedier solution in all but the simplest of cases.

2.3.4 MEASURES OF DISTRIBUTION SHAPE

Along with the use of measures of average and variability to describe a frequency distribution the coefficients of skewness or kurtosis, which refer to the shape of the distribution, may be calculated.

(*a*) *Coefficient of skewness.* In Section 2.3.1 a skewed distribution was introduced and illustrated on Figure 2.17 which was a positively skewed distribution since the tail of the distribution was to the right of the mode. A measure of the degree of skewness of a distribution is given by a dimensionless quantity termed the coefficient of skewness, usually denoted by β_1 where:

$$\beta_1^2 = \frac{\mu_3^2}{\mu_2^3} \qquad \begin{bmatrix} \beta = \text{Greek } beta \\ \mu = \text{Greek } mu \end{bmatrix}$$

μ_r is the mean of the rth power of the deviations from the true mean, and is usually termed the rth moment about the mean. Thus, μ_2 is the 2nd moment about the mean, μ_3 is the 3rd moment about the mean.

$$\therefore \mu_2 = \frac{\Sigma f_i(x_i - m)^2}{\Sigma f_i} \quad \text{(i.e. the variance)}$$

and

$$\mu_3 = \frac{\Sigma f_i(x_i - m)^3}{\Sigma f_i}$$

Three further measure of skewness sometimes used are Pearson's first and second coefficients of skewness and the quartile coefficient of skewness. They are defined as:

$$\text{Pearson's 1st coefficient of skewness} = \frac{\text{mean-mode}}{\text{standard deviation}} = \frac{m\text{-mode}}{s}$$

$$\text{Pearson's 2nd coefficient of skewness} = \frac{3(\text{mean-median})}{\text{standard deviation}} = \frac{3(m\text{-median})}{s}$$

$$\text{The quartile coefficient of skewness} = \frac{(Q_3 - Q_2) - (Q_2 - Q_1)}{Q_3 - Q_1}, \text{ where } Q_1,$$

Q_2, and Q_3 are the first, second, and third quartiles.

The coefficient of skewness is zero for a distribution which is symmetrical about its mean. If the distribution is skewed to the right the coefficient of skewness is positive; if it is skewed to the left, then the coefficient of skewness is negative.

(*b*) *Coefficient of kurtosis.* Kurtosis is a measure of the degree of peakiness of a uni-modal distribution. The coefficient of kurtosis which is a dimensionless quantity and usually denoted by β_2 is:

$$\beta_2 = \frac{\mu_4}{\mu_2^2}$$

where μ_4 is the 4th moment about the mean and μ_2 is the variance.

For a normal distribution (*see* Section 3.4.2) the coefficient of kurtosis = 3. If the coefficient of kurtosis for a distribution is less than 3, then the distribution is less peaked than a normal distribution and is termed platykurtic; if the coefficient of kurtosis is greater than 3, the distribution is more peaked than the normal and is termed leptokurtic. Figure 2.21 illustrates these points.

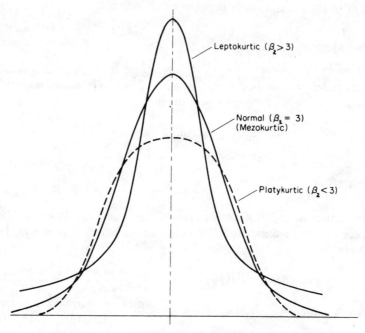

Figure 2.21   Kurtosis

2.4 SUMMARY

The reduction of raw data to more compact and comprehensible forms has been illustrated with reference to graphical representation and the use of statistics.

The two types of graphs usually used are *bar charts* and *frequency histograms*—the bar chart being used for ungrouped discrete variables and the histogram for grouped discrete or continuous variables. On occasions other forms of graphical representation of data which may be used include the *cumulative frequency histogram* (or *ogive*), *pie chart*, *pictograph*, and *Z-chart*.

Statistics used to represent data are generally divided into those which give a measure of average (central tendency) such as the *mean, median,* and *mode*, and those which measure spread, such as *range, variance,* and *standard deviation*. Two further statistics which are sometimes used to measure variability are the *coefficient of variation* and *quantiles*. If it is required to measure the shape of a distribution, then the *coefficients* of *skewness* and *kurtosis* may be calculated.

EXERCISES

1. This table is taken from the records of a local authority's Further Education Centre. Draw a graph of the figures and comment briefly on it.

Attendance at Further Education Class 1968–69

Autumn Term		Spring Term		Summer Term	
Week Number	*No. of Students*	*Week Number*	*No. of Students*	*Week Number*	*No. of Students*
1	20	1	15	1	10
2	22	2	14	2	12
3	22	3	14	3	10
4	18	4	16	4	11
5	17	5	13	5	5
6	19	6	10	6	9
7	15	7	12	7	8
8	13	8	12	8	11
9	15	9	13	9	8
10	12	10	14	10	6

(*IOS, Part 1*)

2. Given below are two sets of data which show the number of coughs given by individual animals, some treated with a drug and some not, when they are placed in an irritant atmosphere. We have reason to believe that the two sets of data might be different from each other and to examine this we wish to draw a histogram for each set. Construct the two frequency tables suitable for drawing the histograms.

(a)	18	18	8	14	18	12	16	9	28	20
	19	17	21	32	20	11	25	13	8	21
	15	16	15	7	17	19	11	7	8	17
	32	21	30	13	23	18	14	9	20	32
	19	17	10	29	17	14	25	22	17	10
	17	31	23	10	18	15	16	0	15	12
	5	20	12	14	14	9	18	5	13	19
	22	15	10	20	26	17	12	15	17	12
	19	20	17	19	11	20	23	9	24	16
	14	11	26	11	35	18	11	11	6	18
	3	18	20	18	12	9	22	14	21	7
	10	9	10	10	11	0	20	14	8	19

(b)	27	23	6	1	1	0	3	20	12	13
	12	20	20	0	21	20	10	32	12	15
	18	16	35	20	15	18	19	20	27	27
	30	39	28	30	14	25	26	5	18	22
	7	16	9	12	16	12	27	28	33	28
	7	22	28	14	15	20	17	33	0	2
	13	18	7	5	0	3	22	23	13	10
	8	25	11	13	2	3	2	15	16	1
	11	30	14	16	8	6	1	12	17	35
	30	18	22	24	15	20	16	19	25	24
	2	9	13	7	26	13	30	43	28	6
	8	40	13	9	3	4	4	10	0	7

(*From IOS, Part 1*)

3. A market research bureau asked 100 housewives who purchased Brand *X* washing powder how many packets they bought during a specified period. The purchases recorded were as follows:

1	1	3	3	3	9	9	20	2	22
10	16	1	46	11	10	24	10	13	1
22	2	14	1	7	6	1	1	6	20
15	3	2	3	23	6	15	3	15	1
7	1	1	1	1	4	1	10	2	30
6	17	5	22	2	23	10	22	3	6
1	3	2	3	1	3	23	7	24	1
3	1	1	1	1	1	14	4	11	8
4	4	9	1	6	1	6	1	13	2
10	6	10	9	4	2	16	1	3	9

Arrange these data into an appropriate frequency distribution and calculate the mean from the grouped data.

Draw a suitable diagram to illustrate the frequency distribution.

(*IOS, Part 1*)

4. Construct histograms to illustrate the following unemployment figures:

Duration of unemployment in weeks	Males	Females
One or less	46,675	13,202
Over 1 and up to 2	37,606	9,229
„ 2 „ „ 3	23,671	4,935
„ 3 „ „ 4	22,107	3,941
„ 4 „ „ 5	20,422	3,774

Exercise 4—*contd.*

Over 5 and up to 6	19,320	3,612
„ 6 „ „ 7	18,542	3,440
„ 7 „ „ 8	15,237	3,086
„ 8 „ „ 9	14,555	2,727
„ 9 „ „ 13	47,046	9,186
„ 13 „ . „ 26	80,893	13,344
„ 26 „ „ 39	40,144	4,986
„ 39 „ „ 52	25,424	3,002
Over 52	82,534	8,308
Total	494,176	86,772

Source: Employment and Productivity Gazette.

(*IOS, Part 1*)

5. The following grouped frequency distribution describes the measurements of content of 200 containers (in cm³).

(*a*) Draw the histogram for these data and thus determine the modal volume.

(*b*) Draw the cumulative frequency curve and estimate from it the median value of volume.

Volumes of containers in cm³	*Number of containers*
6 and less than 7	3
7 „ „ 8	6
8 „ „ 9	49
9 „ „ 10	121
10 „ „ 11	19
11 „ „ 12	2
Total	200

(*IOS, Part 1*)

6. The produce of British farms and market gardens is worth about £2,000 million/year. More than two-thirds of farm sales comes from animals and animal products, less than one-third from crops. The table below shows the 'make-up', by value, of total production in 1967–68.

United Kingdom Farm Sales, 1967–68
(total value—£1,933 million)

Product	*Percentage of sales*
Beef and veal	16·0
Pork and bacon	10·8
Mutton and lamb	4·4

United Kingdom Farm Sales, 1967–68—*contd.*

Product	Percentage of sales
Milk and milk products	22·7
Eggs	8·9
Poultry	5·1
Grain	12·4
Other farm crops	7·5
Horticulture	10·0
Other products	2·2

Draw a circular or 'pie' chart to illustrate the data in the above table.

(*IM, Part 2*)

7. From the following table represent the growth of each type of medium by means of multiple bar charts.

Post-war growth in advertising, by media
£ million (current prices)

Media	1948	1955	1965
Press	65	148	282
Television	—	2	106
Catalogues and free samples	26	53	65
Poster and transport	11	15	18
Outdoor signs	3	11	15
Miscellaneous	16	48	104
Total	121	277	590

(*IM, Part 2*)

8. Below are monthly production figures, in '000 units, of an industrial company.

Monthly Production of the 'XL' Company, 1967/69
('000 Units)

	Jan.	Feb.	Mar.	Apr.	May.	Jun.	Jul.	Aug.	Sep.	Oct.	Nov.	Dec.
1967	400	410	360	400	420	450	430	380	410	450	460	430
1968	420	430	410	380	410	470	450	400	460	490	510	480

Calculate and plot a Z chart for the year 1968.

(*IM, Part 2*)

9. Calculate the mean, median, and mode of the following data. Which average do you consider to be the most useful and why?

Mangolds: Estimated Yield per 5,000 square metres

County	Tonnes	County	Tonnes
Bedfordshire	32·7	Derbyshire	20·9
Berkshire	24·4	Devon	27·1
Buckinghamshire	31·5	Dorset	34·5
Cheshire	18·5	Durham	20·9
Cornwall	26·0	Essex	24·2
Cumberland	19·0	Kent	25·6

From Agricultural Statistics 1963/64

(*IOS, Part 1*)

10. Calculate the mean, median, and mode for the following distribution of scores:

Score x	10	11	12	13	14	15	16	17	18	19	20
Frequency f	2	4	5	8	10	16	10	9	6	4	1

Compare the advantages and disadvantages of the mean and median as measures of central tendency.

(*IOS, Part 1*)

11. Calculate the arithmetic average, the standard deviation, and the coefficient of variation from the following data.

Concentrations of solution (g/l.)

14·2   15·6   13·7   12·9   13·4   13·6   14·0   15·1   14·5   15·0

Explain briefly the uses of the coefficient of variation.

(*IOS, Part 1*)

12. (*a*) The arithmetic mean, and median are two measures of the average value of a set of data. Define these two measures for grouped data and calculate the median and mean for the following data representing the IQ of 100 children at a Junior School:

IQ	No. of children with given IQ
50–59	1
60–69	2
70–79	8
80–89	18
90–99	23
100–109	21
110–119	15
120–129	9
130–139	3

(*b*) Using the formula for the standard deviation of grouped data calculate this quantity for the data given in part (*a*).

(IOS, Part 1)

13. The results of an aptitude test taken by 145 people are tabulated as follows:

Score	Frequency
9	5
8	10
7	19
6	23
5	35
4	21
3	15
2	11
1	6

Calculate the variance:

    (*a*) Using the calculated sample mean.
    (*b*) Using an assumed mean of 3.

(IOS, Part 1)

14. Calculate:

    (*a*) the arithmetic average, and
    (*b*) the inter-quartile deviation

of the following distribution:

Age Distribution of Employees of a Retail Store

Ages			No. of employees
16 years and under 21			16
21 ,, ,, 26			32
26 ,, ,, 31			111
31 ,, ,, 36			215
36 ,, ,, 41			378
41 ,, ,, 46			300
46 ,, ,, 51			242
51 ,, ,, 56			15
		Total	1,309

(IM, Part 2)

15. (*a*) Write a brief description of the 'Ogive' and state where this can be used to the greatest advantage.

(*b*) Draw a cumulative frequency curve of the data given below and estimate the median and quartiles.

Age Distribution of Employees in a Factory, January 1969

Age group	No. of employees
Under 20 years	98
20 years and under 25 years	153
25 ,, ,, 30 ,,	231
30 ,, ,, 35 ,,	276
35 ,, ,, 40 ,,	260
40 ,, ,, 45 ,,	215
45 ,, ,, 50 ,,	147
Total	1,380

(*IM, Part 2*)

16. The following data show the numbers of packets of washing powder purchased by 100 housewives during one month. Construct an ogive from the figures and hence estimate the median purchase.

No. of packets purchased	No. of housewives
X	F
1–3	40
4–6	19
7–9	12
10–12	11
13–15	9
16–18	5
19–21	2
22–24	2
Total	100

(*IOS, Part 1*)

17. From the data given below, draw an ogive, and from it estimate:

(*a*) the median and semi-interquartile range,
(*b*) the percentage of the population in the age ranges 10–19, 20–29, and 30–39.

Check the answers to (*a*) by calculation.
The data represent a standardized percentage age distribution for established towns.

Age	% of total
0–4	8·2
5–14	12·9
15–24	14·1
25–34	15·3
35–44	15·7
45–54	13·0
55–64	10·3
65–69	4·1
70 and over	6·4
	100·0

(*IOS, Part 1*)

18. What is meant by 'skewness'? How is it measured? Explain the meaning of negative and positive skewness. Calculate the quartile coefficient of skewness of the following frequency distribution:

Weight	Frequency (Hundreds)
Under 100	1
100–109	14
110–119	66
120–129	122
130–139	145
140–149	121
150–159	65
160–169	31
170–179	12
180–189	5
190–199	2
200 and over	2

(*IM, Part 2*)

3. *Measuring Uncertainty*

3.1 UNCERTAINTY AND PROBABILITY

In the introductory chapter the phrase 'management gamble' was used to characterize the usual management decision-making situation of having incomplete information about the problem and hence being *uncertain* of the consequences following a given decision. '*Uncertain*' is deliberately used rather than '*ignorant*' because the information available, incomplete though it may be, is usually sufficient to make some consequences seem more *probable* than others.

Once it is appreciated that uncertainty and probability are opposite faces of the same coin it becomes reasonable to examine the formal results of probability theory for their relevance to management decision-making.

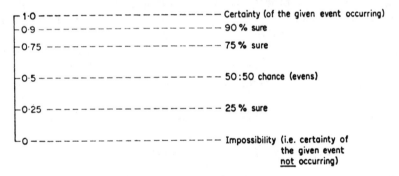

Figure 3.1 Scale of probability

Although 'probability' is a concept in everyday use attempts to define the concept rigorously quickly lead to philosophical complications. The following points are, however, sufficient to define the view of probability which will be taken in this book:

1. A statement about the probability of occurrence of an event is a statement about one's strength of belief that the event will occur.

2. Probability (strength of belief) can conveniently be measured on a scale ranging from 0 to 1·0 (Figure 3.1).

55

On this scale 0 indicates certainty that the designated event will *not* occur, 1·0 indicates certainty that it *will* occur and 0·5 indicates 'perfect' uncertainty as to whether or not it will occur.

For some purposes it is useful to change the scale from 0–1·0 to 0–100%. An example of this is in the specification of significance level (Chapter 6).

3. The usual notations for expressing probabilities takes the form $P(A) = 0.75$ (say) and $P(k = 0) = 0.32$ (say). These are read as: the probability of occurrence of event A is 0·75 and the probability that a variable k equals 0 is 0·32, respectively.

The most common variants of this notation are the immediately obvious elaborations $Pr(A)$ and $Prob(A)$.

Difficulties arise in obtaining numerical values of the type specified. The most straightforward method is that of equating the probability of occurrence of event A with the *relative frequency* of occurrence of event A when a large number of observations have been made. This is of course an *empirical* method but is particularly appropriate to a book focusing on the analysis of data. Readers interested in other aspects of probability are referred to Lindley [1] or Savage [2] where a very different approach is discussed.

The reader is asked to note that while 'strength of belief' is a perfectly general interpretation of any probability measure the interpretation of a probability measure as an expected relative frequency of occurrence is only appropriate where the idea of repeated observations can be given meaning.

Consider the proposition: 'Loch Ness contains a monster'. One can say that the probability of this being true is 0·60, indicating one's strength of belief in the truth of the proposition. There is no natural relative frequency interpretation of this probability statement. This difficulty is discussed from a slightly different angle in connection with Confidence Intervals (Section 4.5.1).

3.2 RELATIVE FREQUENCY HISTOGRAM

In this section the mathematical representation of uncertainty will be illustrated by means of *relative frequency* histograms, which are natural forerunners to *probability distributions*, the more usual method of representing uncertainty.

The relative frequency of occurrence of any event is simply the actual frequency of occurrence of the event divided by the total number of observations; the sum of the relative frequencies of all possible events equalling 1·00. This is illustrated using the distance data of Table 2.5 by constructing a relative frequency histogram where the *areas* of the rectangles of the histogram equal relative frequencies.

The relative frequencies of the data of Table 2.5 are given in Table 3.1, from which it is seen that their sum equals 1·0.

A common definition of probability is the relative frequency definition which states that the probability of an event is equal to the relative frequency of occur-

rence of that event in a large number of trials. Thus, from Table 3.1 one estimates that the probability of a salesman travelling between 125 and 149 kilometres in one week is 0·130; that the probability of a salesman travelling between
150 and 174 kilometres is 0·250, etc. From the information available the probability of a salesman travelling between 25 and 324 kilometres in one week is 1,
i.e. all the salesmen in the sample travel a distance somewhere between 25 and

TABLE 3.1

Relative frequencies (car distances)

Class	Frequency	Relative frequency
0–24	0	0/200 = 0·000
25–49	1	1/200 = 0·005
50–74	2	2/200 = 0·010
75–99	4	4/200 = 0·020
100–124	13	13/200 = 0·065
125–149	26	26/200 = 0·130
150–174	50	50/200 = 0·250
175–199	40	40/200 = 0·200
200–224	30	30/200 = 0·150
225–249	20	20/200 = 0·100
250–274	10	10/200 = 0·050
275–299	3	3/200 = 0·015
300–324	1	1/200 = 0·005
325–349	0	0/200 = 0·000
Total 200		1·000

324 kilometres and the sum of the relative frequencies and therefore of the
probabilities equals 1.

Before plotting a relative frequency histogram it is necessary to determine the
heights of the rectangles. Remembering that for a rectangle:

area = base × height
∴ height of rectangle = area/base.

For the first class in Table 3.1 the area = 0, the base = 25, therefore the
height = 0/25 = 0.

For the second interval of Table 3.1 the area = 0·005, the base = 25, therefore
the height = 0·002.

Proceeding in this way Table 3.2 can be obtained which gives the height of the
rectangle for each class. Once these heights have been obtained the relative
frequency histogram can be constructed as in Figure 3.2.

Given a relative frequency histogram, estimated probabilities can be obtained

TABLE 3.2

Heights of histogram rectangles

Class	Height of rectangle
0–24	0·000/25 = 0·0000
25–49	0·005/25 = 0·0002
50–74	0·010/25 = 0·0004
75–99	0·020/25 = 0·0008
100–124	0·065/25 = 0·0026
125–149	0·130/25 = 0·0052
150–174	0·250/25 = 0·0100
175–199	0·200/25 = 0·0080
200–224	0·150/25 = 0·0060
225–249	0·100/25 = 0·0040
250–274	0·050/25 = 0·0020
275–299	0·015/25 = 0·0006
300–324	0·005/25 = 0·0002
325–349	0·000/25 = 0·0000

from areas of the histogram. For example, the estimated probability of a salesman travelling between 200 and 224 kilometres in one week is equal to the area of the rectangle constructed on a base whose class mark is 212 and width 25, thus:

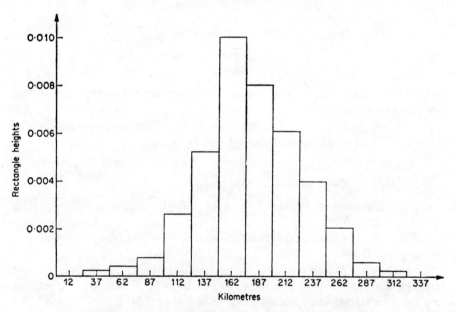

Figure 3.2   Relative frequency histogram (car distances)

Probability = area = base × height
$$= 25 \times 0.006$$
∴ probability = 0·150

The estimated probability of a salesman travelling between 100 and 149 kilometres in one week is the area under the two rectangles whose class marks are 112 and 137. That is:

Probability = (25 × 0·0026) + (25 × 0·0052)
$$= 0.195$$

The estimated probability of a salesman travelling between 25 and 324 kilometres in one week is:

Probability = (25 × 0·0002)+(25 × 0·0004). . .+(25 × 0·006)+(25 × 0·002)
$$= 1.0$$

as would be expected since on the basis of the information on which the histogram is constructed it is certain that a salesman must travel somewhere between 25 and 324 kilometres in one week.

Note that the above probabilities can be obtained directly from the relative frequency table. The purpose of the above calculations is simply to demonstrate that estimated probabilities can be obtained from areas of a relative frequency histogram and that the total area under the histogram equals 1·0. Once continuous probability distributions are covered it will be seen that probabilities can be obtained from areas under curves and the curves must be constructed in such a way that the area under such curves equals 1·0.

3.3 PROBABILITY DISTRIBUTIONS—CONTINUOUS

3.3.1 PROBABILITY DENSITY FUNCTION

To make probability statements about a continuous variable a *probability density function* (p.d.f.) is generally used rather than the relative frequency histogram discussed in Section 2.2.2(*b*).

A p.d.f. is illustrated on Figure 3.3 where a smooth curve is superimposed on the relative frequency histogram of Figure 3.2. The equation of the smooth curve is termed a probability density function and the area enclosed by the curve is unity.

There are two main reasons for using probability density functions (which can take different shapes) in preference to histograms.

1. They are more manageable from a mathematical point of view. This advantage will become apparent later.

2. Due to the finiteness of a sample, histograms are subject to fluctuation and consequently are unreliable for making probability statements about a variable. This may be illustrated as follows. A histogram is a representation of the data obtained from a *sample*, which may be defined as a finite set of items drawn from a *population*. For example the samples may be 'number of kilometres travelled/week by 100 salesmen', 'number of litres of petrol bought by customers covered in a survey of Brand *X* purchasers'. The respective populations, which we defined as all items having a given characteristic could be 'all the salesmen in the

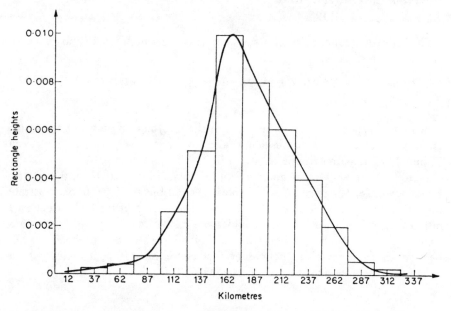

Figure 3.3   Continuous relative frequency distribution (probability density function)

organization', 'all purchasers of Brand *X* petrol'. More will be said about samples and populations in Chapter 4. A relative frequency histogram is a finite sample size estimate of what the underlying distribution of a continuous variable is conceived to be. The p.d.f. may be thought of as the limit in infinite samples of the histogram with very small class intervals.

There are a number of probability density functions which have standard mathematical formulae and it is particularly convenient to use one of these standard forms if the raw data can be shown to be reasonably well represented by it. In the following section a p.d.f. termed the *normal distribution* will be discussed in detail. Probabilities are obtained from p.d.f.s in the same way as from relative frequency histograms—the probability of a variable taking a value between any two limits is equal to the area under the curve between these limits.

3.3.2 THE NORMAL (OR GAUSSIAN) DISTRIBUTION

The *normal distribution* (sometimes referred to as the Gaussian distribution or curve) is a probability density function which has many uses in statistical work, particularly in relation to sampling (*see* Chapter 4).

The normal distribution graphs as a bell-shaped curve which is symmetrical about its mean value and it has certain important properties with respect to its

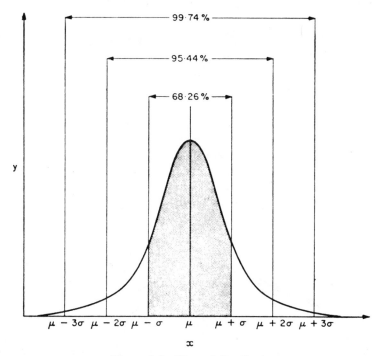

Figure 3.4   Normal distribution

standard deviation. The normal distribution is illustrated on Figure 3.4 and the equation of the curve is:

$$y = \frac{1}{\sigma\sqrt{2\pi}}\, e^{-(x-\mu)^2/2\sigma^2}$$

where: σ (Greek lowercase *sigma*) = standard deviation of the distribution
   σ^2 = variance of the distribution
   π = 3·1419 . . . $\doteq$ 22/7
   e = 2·71828 . . .
   μ = mean of the distribution (i.e. of the *x*s)
   x = variable

For the normal distribution the following approximations are useful to remember:

about 68% of all values of the variable are within the range $\mu \pm 1\sigma$
„   95%   „   „   „   „   „ $\mu \pm 2\sigma$
„   99%   „   „   „   „   „ $\mu \pm 3\sigma$

Considerable use of the above properties is made in the following chapter on sampling and again in a later chapter with reference to significance testing.

Note that in the equation of the normal curve the symbols μ and σ have been used rather than m and s.

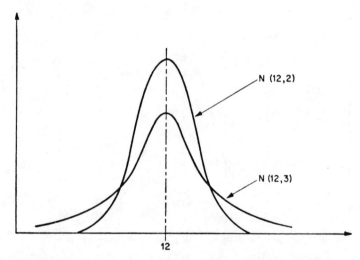

Figure 3.5   Normal distribution—identical means, different standard deviations

In statistical work m and s are usually used for the mean and variance of a *sample*. As will be seen in the following chapter inferences are made about the mean and standard deviation (μ and σ) of the parent population on the basis of the sample mean and standard deviation (m and s). Probability calculations based on p.d.f.s are always obtained from population statistics which are usually inferred from sample statistics. More will be said about this topic in the following chapter.

Before proceeding with probability calculations based on the normal distribution the fact that the standard deviation is a measure of dispersion can easily be demonstrated with reference to Figure 3·5. There a normal distribution has been drawn for two distributions each with a mean of 12 units but one has a standard deviation of 2 units while the other has a standard deviation of 3 units. The usual way of specifying the distribution is:

$N(12, 2)$ and $N(12, 3)$

where N stands for normal, the first figure within the brackets refers to the mean and the second refers to the standard deviation. In some texts the notation N(mean, variance) is used rather than N(mean, standard deviation) and care must be taken to identify which convention is used. The convention N(mean, standard deviation) will be used throughout this book.

Figure 3.5 illustrates that the $N(12, 3)$ distribution has greater spread than the $N(12, 2)$ distribution. Since the spread is greater for the $N(12, 3)$ distribution the height of the distribution at the mean value of 12 must be less than that for the $N(12, 2)$ distribution because the area under each curve or total probability must equal unity.

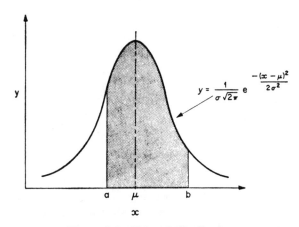

Figure 3.6　$N(\mu, \sigma)$ distribution

The usual method of obtaining areas under curves is to resort to calculus and integrate. Thus, the probability of a value between two general points, a and b on a $N(\mu, \sigma)$ distribution is:

$$\frac{1}{\sigma\sqrt{2\pi}}\int_a^b e^{-\frac{(x-\mu)^2}{2\sigma^2}}\,dx$$

which is a rather awkward calculation. This required probability is illustrated graphically by the shaded area in Figure 3.6.

Because the normal distribution occurs under many different circumstances and because it is used extensively in statistics, tables have been constructed which enable probabilities to be obtained rapidly without resorting to integration. Such tables are given in Table T.1 (page 236).

The tables of the normal distribution are based on what is termed a standardized normal distribution, whose mean is 0 and standard deviation is 1. It is relatively easy to obtain probability information about any normal distribution from the standardized normal distribution, and vice versa. If a variable X is distributed normally with a mean of μ and standard deviation equal to σ, then the variable $(X - \mu)/\sigma$ is distributed normally with mean of 0 and standard deviation equal to 1.

To illustrate the use of tables of the normal distribution consider the following example: Suppose that the distribution of invoice values for sales of a company

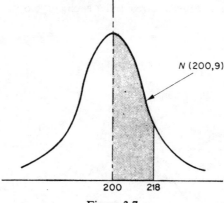

Figure 3.7

is normal, with a mean of £200 and a standard deviation of £9, i.e. it is $N(200, 9)$. What is the probability of one invoice taking a value between £200 and £218? The required probability is represented pictorially by the shaded portion of Figure 3.7.

To obtain this required probability the infor.nation of Figure 3.7 must be converted in such a way that the answer can be obtained from Table T.1.

The standardized normal distribution is illustrated on Figure 3.8 and this effectively is what is given in the tables where probabilities are given for values of z from 0 upwards. The shaded portion of Figure 3.7 is equivalent to the shaded portion of Figure 3.8. The values of z that make the two shaded proportions equivalent are obtained as follows.

The mean value 200 of Figure 3.7 is equivalent to the mean of 0 of Figure 3.8. Tables of the standardized normal distribution are based on what is termed values of deviations from the mean, in units of standard deviations. In other words, z_1 is:

$$z_1 = \frac{\text{deviation of 218 from its mean}}{\text{standard deviation of relevant distribution}}$$

$$= \frac{218-200}{9} \quad = \frac{18}{9}$$

$$= 2$$

From Table T.1, when $z = 2$, the probability $= 0 \cdot 4772$. This is then the probability of an invoice having a value of between £200 and £218.

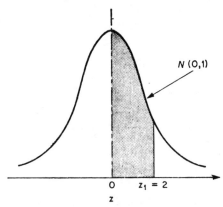

Figure 3.8

Note that for the normal distribution the areas on either side of the mean equal $0 \cdot 5$, hence it is only necessary in the tables for positive values of z to be given. For negative values of z the required probabilities can still be obtained from the tables due to the symmetrical nature of the normal distribution.

For example, if for the $N(200, 9)$ distribution given previously the probability of an invoice value between £191 and £200 was required the shaded area of Figure 3.9 would represent this probability.

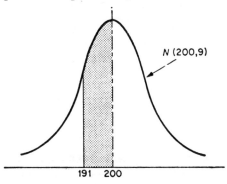

Figure 3.9

The equivalent points on the $N(0, 1)$ distribution are 0 and -1. This is illustrated on Figure 3.10.

Although a probability for a value between 0 and $z_1 = -1$ cannot be obtained from the Table T.1 it is seen from Figure 3.10 that the area between -1 and 0 is

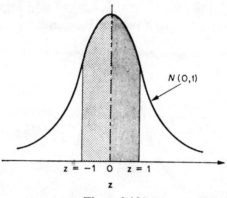

Figure 3.10

identical to that from 0 to $+1$. Thus the required probability of 0·3413 can be obtained from the table for a value of $+1$. This, of course, is also the required probability for $z_1 = -1$.

As a further example consider the question: What is the probability of a sales invoice having a value between £209 and £218?

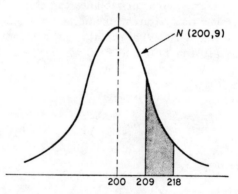

Figure 3.11

Pictorially the shaded area of Figure 3.11 represents the required probability. This is equivalent to Figure 3.12, where the values of z_1 and z_2 are obtained from 209 and 218 respectively, and the standard deviation of 9.

$$z_1 = \frac{209 - 200}{9} = 1$$

$$z_2 = \frac{218 - 200}{9} = 2$$

From tables, when $z = 2$ the probability between 0 and 2 is 0·4772; therefore, the required probability is 0·4772 − 0·3413 which equals 0·1359.

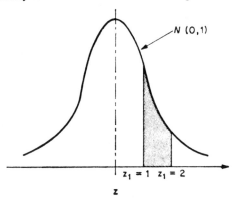

N (0,1)

$z_1 = 1$ $z_1 = 2$

z

Figure 3.12

If the question had been: 'What is the probability of an invoice having a value between 212 and 218?' then the required z values would be:

$$z_1 = \frac{212 - 200}{9} = 1.33$$

$$z_2 = \frac{218 - 200}{9} = 2.00$$

∴ the probability of an invoice value being between 212 and 218 is 0·4772 − 0·4082 = 0·069 (= 6·9%).

To illustrate a further use of tables of the normal distribution consider the following question. If the invoice values of a particular item are normally distributed with a mean of £200 and standard deviation of £9 what is the probability of an invoice having a value greater than £217?

The required probability is illustrated on Figure 3.13 and the appropriate z value is:

$$z = \frac{217 - 200}{9} = 1.89$$

Table T.1 gives the probability of an invoice taking a value of between 200 and 217 as 0·4706. However, since there is a probability of 0·5 of an invoice having a value greater than 200 in the right-hand side of Figure 3·13 the required probability is 0·5 − 0·4706 = 0·0294.

In this section the only continuous distribution discussed has been the normal distribution. There are other continuous distributions but since they are less often useful in marketing studies they have been omitted from this text. There are, however, probability distributions for discrete variables which are used in marketing studies and they are covered in Section 3.6.

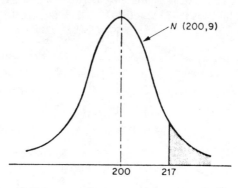

Figure 3.13

3.4 ARITHMETIC OF PROBABILITY

The probability of occurrence of any single event, $P(A)$, say, can generally be interpreted as the relative frequency of occurrence of that event, and, conversely, the relative frequency of occurrence of a single event is an estimate of its probability of occurrence.

Example. If the probability of a television set failing within its period of guarantee is given as 0·06, then it is expected that 6% of similar television sets will fail within the period of guarantee.

Often the probabilities of single events are known or can be estimated but interest is focused on complex events which are combinations of these single events and whose probabilities of occurrence cannot be determined readily.

Example. The probability of any single television set failing within the guarantee period is known. If a dealer sells 6 sets in a week what is the probability that more than half of these sets will be returned within the guarantee period?

Probabilities of complex events are calculated by combining probabilities of simple events by multiplication and/or addition and subtraction. The *arithmetic of probability* prescribes the rules according to which such operations can be performed.

3.4.1 MULTIPLICATION

Suppose that two market surveys are being undertaken; one involves establishing the incidence of hay-fever in a given population and the other the proportion of Brand X cigarette purchasers in the same population. Both surveys are conducted by interviewing passers-by stopped at random on a street corner. Both interviewers agree to question the same 100 people. The results are as tabulated below:

Hay-fever	*Brand X Cigarettes* Purchasers	Non-purchasers	*Total*
Sufferers	5	15	20
Non-sufferers	20	60	80
Total	25	75	100

(Note this simultaneous classification of a number of observations according to two criteria is called a two-way contingency table.)

Using the relative frequency interpretation of probability it is easily seen that this leads to the estimates:

P(passer-by is Brand X purchaser) $= 25/100 = 0.25$
P(passer-by is hay-fever sufferer) $= 20/100 = 0.20$

Similarly it can be seen in this case that:

P(passer-by is a Brand X purchaser *and* hay-fever sufferer) $= 5/100 = 0.05$

This last probability is of course that of a complex event composed of the joint occurrence of both the single events specified. It is apparent that as $0.05 = 0.25 \times 0.20$:

P(passer-by is a Brand X purchaser and a hay-fever sufferer) $= P$(passer-by is a Brand X purchaser) $\times P$(passer-by is a hay-fever sufferer)

Why should this relationship hold in this case?

The logic of the situation can be demonstrated by presenting in stages the consequences of dividing the group of interviewees according to the answers to the survey questions.

Stage 1. One group of 100 interviewees.

Stage 2. The Brand X question subdivides this group into Brand X users, $100\,P_{BX}$ in number, and non-Brand X users, $100\,(1 - P_{BX})$ in number. (P_{BX} is of course used to symbolize the estimated probability of an individual being a Brand X user, or, equivalently the proportion of Brand X users expected in any large group.)

Stage 3. If the tendencies to purchase Brand X and to suffer from hay-fever are *independent*, then the proportion of hay-fever sufferers in both of the Stage 2 groups would be expected to be roughly equal. Hence each of the Stage 2 groups can be subdivided into two groups giving four groups in all. Thus, using P_{HS} as the probability of an individual being a hay-fever sufferer the numbers in each of the resulting groups are:

Number of Brand X purchasers who suffer from hay-fever
$$= 100 \times P_{BX} \times P_{HS}$$
Number of non-Brand X purchasers who suffer from hay-fever
$$= 100(1 - P_{BX})P_{HS}$$
Number of Brand X purchasers who do not suffer from hay-fever
$$= 100 \times P_{BX} \times (1 - P_{HS})$$
Number of non-Brand X purchasers who do not suffer from hay-fever
$$= 100 \times (1 - P_{BX}) \times (1 - P_{HS})$$

These numbers can be transformed into relative frequencies and hence into probability estimates by dividing by the number of individuals in the original group, i.e. 100. This gives:

Probability of an individual being a Brand X purchaser
and a hay-fever sufferer $= P_{BX} \times P_{HS}$
Probability of an individual being a Brand X purchaser
and not a hay-fever sufferer $= P_{BX} (1 - P_{HS})$,
etc.

In each case this gives the probability of a complex event as the product of the probabilities of two simple events.

The general rule for calculating the probability of occurrence of a complex event consisting of the joint (or simultaneous) occurrence of any number of *independent* events $A, B, C, \ldots$, etc., is given by:

$$P(ABC\ldots) = P(A) \times P(B) \times P(C) \times \ldots$$

3.4.2. INDEPENDENCE

The necessity for independence of the single events must be emphasized though the assumption of independence must often be based on common-sense considerations.

For practical purposes the *independence* of two events is an expression of the idea that the occurrence or non-occurrence of one event in no way affects the likelihood of occurrence or non-occurrence of the other. The reader may wish to consider the following examples of *non-independence* and will be able to construct others.

(i) Membership of a trade union and support for a particular political party.

(ii) Readership of certain newspapers and socio-economic class.

(iii) Professional occupation and education level.

Note. Of course it may happen, in another survey or sample, that hay-fever *is* found to be related to smoking.

3.4.3 ADDITION

Suppose a market survey interviewer questions passers-by about the number of brothers and sisters (siblings) each has. 100 interviews produce the following pattern of results:

No. of siblings (n)	0	1	2	3	4	5	6	7	8	9
No. of passers-by claiming n siblings	22	39	16	8	7	4	0	2	1	1
Proportion of respondents	0·22	0·39	0·16	0·08	0·07	0·04	0	0·02	0·01	0·01

The proportions given in the last row are of course estimates of the probability of an individual having n siblings, i.e.

$P(n = 0) = 0·22; P(n = 3) = 0·08$, etc.

If one is interested in the probability of a complex event of the type 'having 2, 3, or 4 siblings', then:

$$P(n = 2 \text{ or } 3 \text{ or } 4) = (16 + 8 + 7)/100 = 0·31$$
$$= P(n = 2) + P(n = 3) + P(n = 4)$$

Addition of probabilities is permissible when the complex event is expressible in words as the probability of occurrence of either *A* or *B* or *C*, etc., *and* where the occurrence of *A* precludes the occurrence of *B* or *C*, etc., and the occurrence of *B* precludes the occurrence of *A* or *C*, etc., and so on. In statistical terms the events *A*, *B*, *C*, etc., must be *mutually exclusive*. In the example used it is obvious that if an individual has a total of 2 siblings, then he does *not* also have a total of 1, 3, 4, 5, 6, 7, 8, or 9 siblings. Another way of looking at the question of *mutual exclusiveness* is to recognize that two events *A* and *B* are mutually exclusive if the probability of the joint occurrence of *A* and *B* is zero, i.e. $P(A \text{ and } B) = 0$.

3.4.4 SUMMARY OF RULES

Probabilities of occurrence of single events may be *multiplied* together to give the probability of the complex event defined as the joint or simultaneous occurrence of the single events if, and only if, these single events are independent.

$$P(A \text{ and } B \text{ and } C \text{ and} \ldots) = P(A) \times P(B) \times P(C) \times \ldots$$

Probabilities of occurrence of single events may be *added* to give the probability of occurrence of a complex event defined as the occurrence of at least one of the simple events being considered if and only if these simple events are mutually exclusive.

$P(A \text{ or } B \text{ or } C \text{ or} \ldots) = P(A) + P(B) + P(C) + \ldots$

3.5 EXPECTED VALUES

Suppose one buys a 10p ticket in a lottery for a prize valued at £10. Suppose further that exactly 500 tickets are sold. Given that the draw is 'fair' it is seen that there are 499 'chances' of getting no return from the 'investment' and 1 'chance' of getting a 'return' of £9·90, i.e. £10 prize less the cost of the ticket.

Using the relative frequency interpretation of probability the balance of advantage of this investment can be expressed as:

$$\text{Probable 'value' of investment} = \frac{499}{500} \text{ (loss of 10p)} + \frac{1}{500} \text{ (gain of £9·90)}$$
$$= 0 \cdot 998 \, (-10\text{p}) + 0 \cdot 002 (+ 990\text{p})$$
$$= -9 \cdot 98\text{p} + 1 \cdot 98\text{p} = - 8 \cdot 00\text{p}.$$

This probable value is formally called the *expected value* of the investment in the lottery ticket and measures the return to the investor averaged over a very large number of repetitions of such an investment.

The expected value of any venture is calculated by enumerating the value of all possible outcomes of the venture, multiplying each value by the corresponding probability of occurrence of that outcome and summing over all the outcomes. This may be expressed algebraically as follows.

If there are n possible values $x_1, x_2, \ldots x_n$, and the probabilities of each of these possibilities are $p_1, p_2, \ldots p_n$, then the expected value of x, denoted $E(x)$ is:

$E(x) = p_1 x_1 + p_2 x_2 + \ldots + p_n x_n$

i.e. $E(x) = \sum_{i=1}^{n} p_i x_i$ or $\Sigma \, p_i x_i$, where the summation covers all possible values of i.

Example. Bloggs' soap powder is sold in 20p packets; 10% of these packets contain a voucher valued at 20p which may be exchanged for cash. What is the expected cost of buying Bloggs' soap powder?

The form of the voucher is such that 10% of customers receive a free packet of Bloggs' soap powder. Hence:

$$\text{Expected cost} = 0 \cdot 90 \, (20\text{p}) + 0 \cdot 10 \, (0\text{p})$$
$$= 18\text{p}$$

The reader will recognize that the concepts of 'expected value' and 'average' are almost indistinguishable in large samples. The formal distinction is that expected values are calculated using all possible values which a variable can possibly take, whereas an average value would correctly be calculated over the values of the variable which had in practice been observed. The relationship between the two concepts is parallel to that between probability and relative frequency in that as the number of observations made increases so the practical distinction between the 'theoretical' concept of probability and the observed relative frequency measure vanishes. Similarly average values calculated over sufficiently large numbers of observations will be indistinguishable from the corresponding 'theoretical' expected value.

3.6 PROBABILITY DISTRIBUTIONS—DISCRETE

The multiplication and addition of probabilities are represented particularly clearly in the formulation of standard expressions to calculate the probabilities of occurrence of specified values of discrete random variables in certain frequently recurring situations.

3.6.1 BERNOULLI TRIALS

A situation frequently encountered is where a 'trial' can result only in success (*S*) or failure (*F*) or where a 'question' can result only in a 'Yes' or 'No' answer. If in addition at each 'trial' or 'question' the probability of a favourable response is constant, the trial is described as a Bernoulli trial.

Example. Tossing a coin where a 'Head' might be deemed a favourable outcome.
Example. In market surveys or opinion polls, providing the question asked requires a 'Yes/No' answer, the asking of such a question can be considered as a Bernoulli trial.

The probability distributions to be described are all based on the assumption of Bernoulli trials. The conditions specified for the addition and multiplication of probabilities are immediately satisfied by such trials because implicit in the definition are the ideas of independence of successive trials and of the outcomes being mutually exclusive. (It is convenient to point out here that where a trial must result in either success or failure these outcomes can be described as *exhaustive* because they exhaust the list of possible outcomes. Such a list in other situations might contain more than two possible outcomes. In any situation where a set of outcomes, O_1, O_2, ... O_n, can be described as mutually exclusive and exhaustive it follows from the addition rule that $P(O_1) + P(O_2) + \ldots + P(O_n) = 1$.)

3.6.2. BINOMIAL DISTRIBUTION

(i) If Brand *X* is purchased by 10% of housewives what is the probability that a group of 5 housewives, picked at random, will contain exactly 2 Brand *X* purchasers?

(ii) If 5% of eggs laid are double-yolked what is the probability of the usual sales pack of half a dozen eggs containing no double-yolked eggs?

Both these examples can be looked on as a series of Bernoulli trials; 5 in the first example and 6 in the second.

This of course assumes that the probability of occurrence of the events in question remains constant from trial to trial. In (i) for instance the occurrence of mothers and daughters in the group of 5 might raise doubts that a mother's preference (or prejudice) could be shared by the daughter and hence that their choices of brand could not be considered as independent. This could of course invalidate the Bernoulli trial assumption. Similarly in (ii) if all eggs in a pack came from the same hen the Bernoulli trial assumption might not be valid. In general terms both examples can be expressed as follows.

A particular outcome has a probability 'p' of occurrence. What is the probability of 'r' such occurrences in 'n' trials?

Example 1. $p = 0.10; r = 2; n = 5.$
Example 2. $p = 0.05; r = 0; n = 6.$

The Binomial distribution provides a formula by which such probabilities may be calculated.

Consider the first example. If the identification of a Brand X purchaser is deemed a success (S), then one of the possible group results of the type required is $S F F S F$.

Because these 5 trials are independent and the probability of success is 0·10 one can write:

$$P(SFFSF) = 0.10 \times 0.90 \times 0.90 \times 0.10 \times 0.90 = 0.00729.$$

In the absence of a *numerical* value for the probability of success one could write:

$$P(SFFSF) = p \times (1 - p) \times (1 - p) \times p \times (1 - p) = p^2 (1 - p)^3$$

Similarly the sequence $SSFFF$ would satisfy the specification. Again this can be written as:

$$P(SSFFF) = 0.10 \times 0.10 \times 0.90 \times 0.90 \times 0.90 = 0.00729 \quad \text{or}$$
$$P(SSFFF) = p \times p \times (1 - p) \times (1 - p) \times (1 - p) = p^2 (1 - p)^3$$

Note that one is calculating the probability of 'r' successes where r is taking in this case the value 2. In the expression derived 'p', the probability of success, is written with a superscript '2' or in different language, 'p' is raised to the power of 2. In the general case the 'superscript' or 'power' of p is always the current value of 'r'.

If all k possible sequences which satisfy the specification can be written down then as $p^2 (1 - p)^3$ is the probability of occurrence of each mutually exclusive sequence, the *sum* of these probabilities is the probability of at least one of these sequences occurring and is given by $kp^2(1-p)^3$. How can the value of k most readily be calculated?

The logic of the problem can be demonstrated in terms of the number of different ways 2 letters '*S*' can be placed in 5 spaces. The first letter *S*, call it S_1 for reference, can be placed in any one of 5 positions. For each such position selected 4 are left vacant, any one of which may be selected for the second *S*, S_2 say. Thus $5 \times 4 = 20$ possible pairs of positions may be filled by 2 *S*s. Consideration of these 20 pairs will show that sequences S_1S_2 and S_2S_1 are counted as different. In problems of the sort considered this applies to every pair of *SS*s, hence the 20 sequences must be divided by 2 to give 10 *different* sequences. k in the problem considered is hence 10.

This calculation is simple for small values of r and n (2 and 5 in the problem considered). A general formula for this calculation is expressed compactly using the mathematician's factorial notation. Here $n!$ (read as 'n factorial') $= n \times (n-1) \times (n-2) \times \ldots 2 \times 1$. Thus $5! = 5 \times 4 \times 3 \times 2 \times 1 = 120$; $2! = 2 \times 1 = 2$.

Note that $0!$ is defined to be 1.

The expression $n!/(n-r)!\,r!$ will be found to give the appropriate value of k in every case. (This expression is sometimes represented as ^nC_r and sometimes as $\binom{n}{r}$.)

In the example considered $5!/3!\,2! = 5 \times 4 \times 3 \times 2 \times 1/3 \times 2 \times 1 \times 2 \times 1 = 10$.

Summary. Given a set of n Bernoulli trials in which the probability of success in any trial is p the probability of exactly r successes in this set of trials is given by

$$P(r) = {}^nC_r p^r (1-p)^{n-r} \equiv \binom{n}{r} p^r (1-p)^{n-r} \equiv \frac{n!}{(n-r)!r!} p^r (1-p)^{n-r}$$

Note that $(1-p)$ is usually represented by q. When this is done the above expressions can be rewritten as:

$$P(r) = \frac{n!}{(n-r)!r!} p^r q^{n-r}$$

If a series of n trials is repeated a very large number of times the number of successes r obtained in each series can take any value in the range 0–n. But the average number of successes per series of n trials is given by np and the variance of the number of successes per series by npq. These results are offered without proof. Thus one speaks of the mean and variance of a binomial distribution being given by np and npq respectively.

Example. In the immediately previous example:

Mean no. of Brand X housewives in a group of $5 = 5 \times 0 \cdot 10 = 0 \cdot 5$

Variance of number of Brand X housewives in a group of $5 = 5 \times 0.10 \times 0.90 = 0.45$

Example. In the second introductory example what are the probabilities of egg boxes of 6 containing 0, 1, 2, or 3 double-yolked eggs? What is the average number and the standard deviation of the number of double-yolked eggs in a pack?

$$P(r = 0) = {}^6C_0(0.05)^0 (0.95)^6 = \frac{6!}{6!0!} \times 1 \times 0.7351 = 0.7351$$

$$P(r = 1) = {}^6C_1(0.05) \times (0.95)^5 = \frac{6!}{5!1!} \times 0.05 \times 0.7738 = 6 \times$$
$$0.05 \times 0.7738 = 0.2321$$

$$P(r = 2) = {}^6C_2(0.05)^2 \times (0.95)^4 = \frac{6!}{4!2!} 0.0025 \times 0.8145 = 15 \times$$
$$0.0025 \times 0.8145 = 0.0305$$

$$P(r = 3) = {}^6C_3(0.05)^3 \times (0.95)^3 = \frac{6!}{3!3!} \times 0.000125 \times 0.857375 =$$
$$20 \times 0.000125 \times 0.857375 = 0.0021$$

Mean number of double yolks $= np = 6 \times 0.05 = 0.30$

Standard Deviation $= \sqrt{npq} = \sqrt{6} \times 0.05 \times 0.95 = \sqrt{0.285} = 0.534$

3.6.3. NORMAL DISTRIBUTION AS AN APPROXIMATION TO THE BINOMIAL DISTRIBUTION

Calculations for small values of n are not difficult but as n increases the arithmetic becomes tedious. In such cases it is often possible to use a normal distribution with mean np and variance npq as an adequate approximation to the required binomial distribution. The justification for this and the minor manipulation necessary can be shown by the following example.

Suppose that one is interested in the number of successes in 10 Bernoulli trials where p has the value 0.4. The probabilities of 0, 1, 2 ... 10 successes can be calculated:

No. of successes

(r)	0	1	2	3	4	5	6	7	8	9	10
$P(r)$	0.0060	0.0403	0.1209	0.2150	0.2508	0.2007	0.1115	0.0425	0.0106	0.0016	0.0001

These probabilities can be represented graphically as shown in Figure 3.14 as a bar chart. A slightly artificial alternative representation is permissible where the probabilities are represented by *rectangles* of *unit* width erected on the *intervals* $(-\frac{1}{2}, +\frac{1}{2})$, $(+\frac{1}{2}, +1\frac{1}{2})$, $(+1\frac{1}{2}, +2\frac{1}{2})$, ... $(+9\frac{1}{2}, +10\frac{1}{2})$ instead of vertical *lines* erected on the *values* 0, 1, 2, ... 10.

This is shown in Figure 3.15 as a 'histogram'.

Also shown on this Figure (as a dotted line) is the normal distribution with same mean and variance.

It is seen that the areas under the normal distribution curve and between each pair of rectangle 'uprights' approximate to the true rectangular area. This is equivalent to stating that the normal distribution probability of a value occurring between $(r - 0.5)$ and $(r + 0.5)$ is approximately equal to the true probability of occurrence of r, which has been artificially distributed over the interval

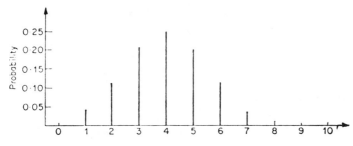

Figure 3.14 Binomial distribution

$(r - 0.5)$, $(r + 0.5)$. In the illustration used the corresponding calculations would be for $r = 2$ and $r = 3$.

Mean $= np = 4.0$. Standard deviation $= \sqrt{npq} = \sqrt{2.4} = 1.55$.
True $P(r = 2) = 0.1209$.

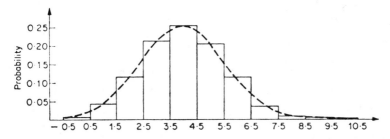

Figure 3.15 Relationship between normal and binomial distributions

Normal distribution probability of a value between 1·5 and 2·5 is obtained from the standard tables using the reference values.

$$z_1 = \frac{4.0 - 1.5}{1.55} = \frac{2.5}{1.55} = 1.61; \quad z_2 = \frac{4.0 - 2.5}{1.55} = \frac{1.5}{1.55} = 0.97$$

Corresponding areas are (from tables) 0·4463 and 0·3340.

The required probability is the difference of these values $= 0.1123$. Similarly for $r = 3$.

True $P(r = 3) = 0.2150$. Normal distribution approx. $= 0.2085$.

The complete picture of probabilities by the two methods is as shown in the following table:

$r = 0$	1	2	3	4	5	6	7	8	9	10
True (Binomial)										
P 0.0060	0.0403	0.1209	0.2150	0.2508	0.2007	0.1115	0.0425	0.0106	0.0016	0.0001
Approximate (Normal)										
P 0.0119	0.0418	0.1123	0.2085	0.2510	0.2085	0.1123	0.0418	0.0100	0.0017	0.0002

Though the probabilities shown correspond tolerably well the example used has been for a rather small value of n. The satisfactoriness of the approximation improves as *n increases* and as *p approaches the value 0·5*. A criterion commonly recommended is to use the normal approximation when the *smaller* of the values np or nq exceeds 5.0.

Example. A marketing manager is seeking to recruit dealers as accredited agents. Experience tells him that 1 in 10 of dealers approached are likely to agree. If he approaches 300 dealers what is the probability that he will succeed in recruiting (a) more than 25 agents, (b) between 25 and 35 agents, and (c) more than 60 agents?

In this case the direct calculations are obviously formidable. $np = 300 \times 0.10 = 30$ so use of the normal distribution approximation is likely to be accurate.

Mean $= 30$. Standard deviation $= \sqrt{27} = 5.2$.

(a) Probability of recruiting more than 25 agents $= 1 -$ Probability of recruiting 25 agents or less.

$$\text{Table look-up value is } z = \frac{30.0 - 25.5}{5.2} = 0.87$$

From tables this gives a probability of 0.3078. This is the probability of occurrence of a value between 25.5 and 30.0. The probability of a value less than 25.5 is therefore $0.5000 - 0.3078 = 0.1922$. Hence the probability required $= 1 - 0.1922 = 0.8078$.

(b) Required probability is of a value between 24.5 and 35.5.

Table look-up values are:

$$z_1 = \frac{30 - 24.5}{5.2} = 1.06; \quad z_2 = \frac{35.5 - 30.0}{5.2} = 1.06$$

Tables give corresponding probability of 0.3554 for $z = 1.06$. Hence required probability $= 0.3554 + 0.3554 = 0.7108$.

(c) Probability of more than $60 = 1 -$ Probability of 60 or less. Probability of 60 or less $=$ Probability of a value less than or equal to 60.5.

Table look up value for values *between* 30·0 and 60·5 is:

$$z = \frac{60·5 - 30·0}{5·2} = \frac{30·5}{5·2} = 5·86$$

Corresponding probability $= 0·4999 \ldots$
Table probability $= 0·5000 + 0·4999 \ldots = 0·9999 \ldots$
Hence required probability $= 1·000 - 0·9999 \ldots \doteqdot 0.$

3.6.4 NEGATIVE BINOMIAL DISTRIBUTION

The binomial distribution enables one to calculate the probability of 'r' successes in 'n' Bernoulli trials and hence the average number of *successes* in 'n' trials. The *negative binomial distribution* enables one to calculate the probability of having to conduct 'n' Bernoulli trials in order to obtain 'r' successes and hence the average number of *trials* necessary to obtain 'r' successes. Note that in the binomial case the number of *trials* is predetermined and the number of successes is the discrete random variable, while in the negative binomial case the number of *successes* required is predetermined and the number of trials is the discrete random variable. Note also that if exactly 'n' trials are required to produce exactly 'r' successes the *last* trial must be a success because the last (or 'rth') success required determines the end of the trials. If for instance one is interested in the probability of 6 trials producing exactly 4 successes, then the only sequences of relevant results are of the type:

SSSFFS, SFSSFS, FSSFSS, SSFFSS, etc.

It will be recognized that such sequences can be described as containing $r - 1$ successes in $n - 1$ trials followed by a successful trial. The probability of $r - 1$ successes in $n - 1$ trials is given by the binomial formula.

$$P((r - 1) \text{ in } (n - 1)) = {}^{n-1}C_{r-1} \, p^{r-1} \, q^{(n-1)-(r-1)}$$
$$= {}^{n-1}C_{r-1} \, p^{r-1} \, q^{n-r}$$

Hence $P(n$ trials result in r successes)

$$= {}^{n-1}C_{r-1} \, p^{r-1} \, q^{n-r} \times p.$$
$$= {}^{n-1}C_{r-1} \, p^r q^{n-r}$$

Example. 30% of housewives are believed to be Brand X purchasers. If this is true what is the probability of having to interview 3, 4 or 5 housewives before encountering exactly 3 Brand X purchasers?

$P(3 \text{ interviews gives exactly 3 successes}) = {}^2C_2(0·3)^3 \, (0·7)^0$
$\qquad\qquad\qquad\qquad\qquad\qquad\qquad = 0·0270$
$P(4 \text{ interviews gives exactly 3 successes}) = {}^3C_2(0·3)^3 \, (0·7)^1$
$\qquad\qquad\qquad\qquad\qquad\qquad\qquad = 0·0567$
$P(5 \text{ interviews gives exactly 3 successes}) = {}^4C_2(0·3)^3 \, (0·7)^2$
$\qquad\qquad\qquad\qquad\qquad\qquad\qquad = 0·0794$

(Where 'success' implies that an interview is with a Brand X purchaser.)

The mean of the binomial distribution (np) gives the average number of 'successes' which will be obtained in 'n' trials. Similarly the variance of the distribution (npq) measures the variability of the number of successes obtained in n trials. In the negative binomial case the mean of the distribution gives the average number of trials and the variance of this distribution measures the variability to be expected in the number of necessary trials. The expressions are:

$$\text{Mean no. of trials} = \frac{r}{p}. \quad \text{Variance of no. of trials} = \frac{rq}{p^2}$$

Example. In the immediately previous example the average number of interviews required to encounter 3 Brand X purchases $= \dfrac{3}{0\cdot3} = 10$.

$$\text{Variance} = \frac{3 \times 0\cdot7}{(0\cdot3)^2} = 23\cdot33$$

The reader should note that in the special case of $r = 1$ the corresponding form of the distribution is referred to as the *geometric* distribution. The names given to distributions are of no assistance in determining the appropriateness of their application and it is much more important that the implicit assumptions of a distribution be remembered rather than its title.

3.6.5 THE 'CIGARETTE CARD' PROBLEM

A probability situation related to that considered in Section 3.6.4 and which recurs with variations as a sales promotional device is the collection of a full set of different give-away coupons in order to qualify for a prize. In acknowledgment of its widespread use by cigarette companies in the 1930s it is referred to here as the 'Cigarette Card' problem. In its basic form it is stated thus:

A full set consists of n different cards. Assuming that each packet bought is equally likely to contain any of the n cards (i.e. no deliberately 'scarce' card is created) how many packets on average must be bought to enable a set to be collected? ('Swopping' is excluded from consideration!) The first *packet* bought always contain a card to be included in the buyer's set as that card is the first *card* of the set.

There are now ($n - 1$) cards required to complete the buyer's set hence the probability that the next card is one to be added to the buyer's set is ($n - 1$)/n. The average number of packets to be bought to obtain one of these is, by the negative binomial distribution, $n/(n - 1)$ because $r = 1$; $p = (n - 1)/n$.

After the second card is added to the buyer's set ($n - 2$) desirable cards remain and the average number of packets to be bought is $n/(n - 2)$; ($r = 1$; $p = (n - 2)/n$).

Adding all these average purchases at each stage gives:

Average total purchases $= 1 + \dfrac{n}{n-1} + \dfrac{n}{n-2} + \dfrac{n}{n-3} + \ldots \dfrac{n}{1}$

This can be rewritten as:

$$= \frac{n}{n} + \frac{n}{n-1} + \frac{n}{n-2} + \frac{n}{n-3} + \ldots \frac{n}{1}$$

$$= n\left(\frac{1}{n} + \frac{1}{n-1} + \frac{1}{n-2} + \ldots 1\right).$$

No simple expression exists for this total but a fair approximation is obtained by writing:

Average total purchases $= n(2 \cdot 303 \log_{10} n + 0 \cdot 58)$

or $= n(\log_e n + 0 \cdot 58)$

Example. How many items must be purchased on average to obtain a set if the set consists of (*a*) 5, (*b*) 10, and (*c*) 20 different cards?

(*a*) $n = 5$

Average total number $= 5\left(\frac{1}{5} + \frac{1}{4} + \frac{1}{3} + \frac{1}{2} + \frac{1}{1}\right)$
$= 5(0 \cdot 20 + 0 \cdot 25 + 0 \cdot 33 + 0 \cdot 50 + 1 \cdot 00)$
$= 5 \times 228 = 11 \cdot 40$

Using the approximation:

Average total number $= 5 \times (\log_e 5 + 0 \cdot 58)$
$= 5 \times (1 \cdot 61 + 0 \cdot 58)$
$= 5 \times (2 \cdot 19) = 10 \cdot 95$

(*b*) $n = 10$

Average total number $= 10\left(\frac{1}{10} + \frac{1}{9} + \frac{1}{8} + \frac{1}{7} + \frac{1}{6} + \frac{1}{5} + \frac{1}{4} + \frac{1}{3} + \frac{1}{2} + \frac{1}{1}\right)$
$= 10 (2 \cdot 93) = 29 \cdot 3$

Using the approximation:

Average total number $= 10(\log_e 10 \times 0 \cdot 58)$
$= 10(2 \cdot 30 + 0 \cdot 58)$
$= 10 \times 2 \cdot 88 = 28 \cdot 8$

(*c*) $n = 20$

In this case the exact calculation is lengthy. Using the approximation:

Average total number $= 20(2 \cdot 99 + 0 \cdot 58)$
$= 20(3 \cdot 57) = 71 \cdot 4$

Two points should be emphasized.

(i) The satisfactoriness of the approximation increases as n increases.

(ii) As n increases the average number of items purchased increases *more* than proportionately, i.e. doubling the number of cards in the set *more* than doubles the average number of purchases required to obtain a complete set.

3.7 SUMMARY

Actions in the real world may result in one of a number of possible consequences and uncertainty may exist as to which particular consequence will result
on a particular occasion. By observing the *relative frequency* of occurrence of
different outcomes estimates are obtained of the *probability* of occurrence of
these different outcomes. These probabilities can then be combined using the
arithmetic processes of addition and multiplication to give the probabilities of
occurrence of complex events consisting of combinations of simple outcomes.
The arithmetic processes are valid under carefully defined but not too restrictive
conditions.

It is sometimes possible and, when possible, usually convenient to use mathematical formulae to represent particular patterns of probabilities which are
frequently recurring. Of these the Normal Distribution and the Binomial Distribution are particularly useful.

Though probability is considered in this chapter in terms of the intuitively
acceptable relative frequency interpretation the reader is asked to note that other
interpretations are possible and sometimes necessary. Passing reference is made
to the interpretation of statements about the probable occurrence of singular
events. Here no *simple* relative frequency interpretation seems appropriate and
the use of a probability value as a measure of strength of belief is more natural.

EXERCISES

1. One thousand 50-kg sacks of coal were weighed to the nearest kilogram and
the following frequency distribution obtained:

Weights			Frequency
45 kg and less than 46 kg			1
46	,,	,,	47 ,,	5
47	,,	,,	48 ,,	15
48	,,	,,	49 ,,	50
49	,,	,,	50 ,,	150
50	,,	,,	51 ,,	272
51	,,	,,	52 ,,	215
52	,,	,,	53 ,,	165
53	,,	,,	54 ,,	85
54	,,	,,	55 ,,	25
55	,,	,,	56 ,,	12
56	,,	,,	57 ,,	5
			1,000

Construct a relative frequency histogram of the above data and from it estimate the proportion of sacks of coal having a weight of at least 52 kilograms.

2. The number of cycles of operation which a machine performed before failure occurred has been noted for 6 months and the following information obtained:

Number of cycles before failure	Frequency
0–49	100
50–99	80
100–149	70
150–199	50
200–249	45
250–299	40
300–349	35

Construct a relative frequency histogram and from it estimate the probability of a failure occurring before 150 cycles of operation have been performed.

3. The life of a special purpose lamp is normally distributed with a mean of 58 hours and a standard deviation of 2 hours. What is the probability of a lamp failing between 58 and 62 hours after it has been put into use?

4. The lengths ordered for $\frac{1}{2}$-cm. stainless steel square bar are normally distributed with a mean of 5·75 metres and standard deviation of 0·9 metres. What proportion of orders have:

(*a*) an order length greater than 6 metres?

(*b*) an order length smaller than 5 metres?

5. The mean daily output of operatives in a spinning factory is normally distributed with mean of 540 metres and a standard deviation of 17 metres. The management are introducing more modern machines but will only train workers whose mean daily output is already 525 metres or more to use them. What percentage of the factory force can expect re-training?

(IOS, Part 1)

6. Evaluate: (*a*) $^3C_1$, $^6C_4$, $^3C_0$, $^{10}C_3$, $^{10}C_7$.

(*b*) $\sum_{i=0}^{3} {}^3C_i$; $\sum_{i=0}^{5} {}^5C_i$.

7. A famous advertising slogan claims that 4 out of 5 housewives cannot distinguish between a particular brand of margarine and best butter. If this claim is valid and 5,000 housewives are tested in groups of 5 how many of these groups will contain 0, 1, 2, 3, 4, and 5 housewives who do distinguish between the two products? Assume that the capacity to distinguish between the two products is randomly distributed so that Bernoulli trial conditions are satisfied at each test.

8. What values are obtained using the normal approximation method for Question 7? Why is the agreement seemingly poor?

4. *Samples and Sampling*

4.1 INTRODUCTION

Limitations of time and/or money frequently make it necessary to take decisions on the basis of a small portion of the total information which could conceivably be collected. The 'portion' is known as a *sample* and the 'total information' as the parent population (or, simply, *population*) of the sample. The term population seems natural in the contexts of opinion polls or market surveys where human beings are associated with each item of information but it is completely general in application and could be used to describe for instance electric motors stocked in a warehouse or the weekly production of cans of beer by a brewery.

Intuitively, and correctly, it is felt that the larger the sample the more reliable will be the inferences that can be drawn from that sample. Constraints of time and money affect both the sample size and the method of selecting the sample. In this chapter some of the considerations relevant to adopting particular sampling schemes are discussed together with the way in which the 'reliability' of sample based inferences may be interpreted.

4.2 RANDOM SELECTION

The basic formulae of sampling theory require the idea of random selection in their derivation. In principle this means that a *sampling frame*, or list of items to be sampled from, must be constructed. When this list, which identifies and locates each item in the population being considered, is established, random number tables (Table T.5) are used to decide which items in the list are to be included in the sample. This is achieved by entering the table at an arbitrary point, and reading off in strict sequence a succession of digits. This sequence is then broken up into k-digit numbers which are used to identify the appropriate items in the sampling frame. This procedure is illustrated in the following example.

Example. Suppose a random sample of 8 jars is to be selected from a store containing 6 gross of such jars in order to provide an average weight of contents. It must be supposed that the 864 jars can be numbered and it is convenient to consider the numbering as from 000 to 863 instead of from 001 to 864. Consult random number tables (e.g. Table T.5). Eight three-digit numbers are required and the table may be entered at any starting point. If the starting point is arbitrarily selected as the first digit in the sixth row and the table is read in rows the sequence of digits runs: 16 22 77 94 39 49 54 43 54 82 17 37 93 23 78 87 35 20. Partitioning this into three-digit numbers gives: 162, 277, 943, 949, 544, 354, 821, 737, 932, 378, 873, 520.

943, 949, 932 and 873 are rejected because they are greater than 863 but the remaining numbers are acceptable and the corresponding jars form a valid *random* sample of 8. Note that it is acceptable to start at the second (or any other) digit in the row and to read off: 622, 779, 439, 495, 443, 548, 217, 379 as the three-digit sequences but it is not acceptable to start 162, 277 and then when 949 is encountered simply to reject the 9 and take the next number as 439. In short the sequence of three-digit *numbers* must be accepted from the tables and *numbers* rejected when they exceed the population bounds. Identical rules apply for selecting 2, 4, 5, etc., digit numbers.

4.2.1 SIMPLE RANDOM SAMPLING

This term is used confusingly in different textbooks to describe either of two distinctly different sampling procedures:

1. The sampling procedure in which, after each random selection is made, the item has its relevant properties noted and is then *replaced* in the population before the next random selection. This means that the population is never depleted and can be treated as a population of infinite size.

2. The sampling procedure in which, after each random selection, the item selected is *not replaced* and the next item selected is hence a random selection from the original population *depleted* by the number of items already selected.

These procedures are generally identified as *sampling with replacement* and *sampling without replacement*.

As a rough guide to the textbook confusion British textbooks tend to use the term simple random sampling to describe sampling with replacement while American textbooks use the term to describe sampling without replacement.

4.3 PROPERTIES OF SAMPLE MEANS

The most basic results of sampling theory consist of statements which can be made about the relationships between the means of random samples and the properties of the population from which the samples are drawn. Specifically, if a population of N items has mean μ and variance σ^2, and random samples of size n

are drawn from that population then the means of these samples have the following characteristics:

1. The mean of the distribution of these sample means is equal to μ, the population mean.

2. The variance of the distribution of these sample means is related to the population variance σ^2 in a manner determined by the sample selection procedure used.

If random sampling is done *with* replacement,

Variance of sample means $= \dfrac{\sigma^2}{n}$

If random sampling is done *without* replacement,

Variance of sample means $= \dfrac{\sigma^2}{n} \cdot \dfrac{N-n}{N-1}$

3. In a large class of circumstances the relative frequency distribution of sample means becomes more and more satisfactorily represented by a Normal distribution as n increases.

Though the theoretical justification of these statements is difficult they can be made plausible in terms of a simple numerical example. The statements are important because they form the basis on which arguments are constructed about the reliability of sample based inferences.

Example. A chain-store sells gift vouchers valued £1, £2, £3, and £6. Demand for each type of voucher is equal, i.e. 25% of sales are for £1 vouchers, 25% for £2, 25% for £3, and 25% for £6 vouchers. Suppose that nothing is known about the distribution of demand and that the average and variance of value of voucher sold has to be inferred from samples of data drawn from sales returns.

It will be convenient to consider the problem independently of its context. The demand pattern can be represented by four tickets marked 1, 2, 3, and 6. Sampling without replacement *would* restrict one to a population of size 4 but sampling with replacement allows the population to be considered as having infinite size and allows samples of any desired size to be drawn. Calculation, as in Chapter 2, gives $\mu = 3 \cdot 0$ and $\sigma^2 = 3 \cdot 5$. All possible samples of sizes 2, 3, and 4 are listed in Table 4.1. Note that it is necessary to specify both type and frequency of occurrence of sample to give a complete listing. The argument is essentially similar to that employed in relation to the Binomial coefficient. Thus in a sample of size 2, a (1, 3) type of sample (i.e. a sample containing one ticket marked '1' and one ticket marked '3') could occur in 2 different ways:

(*a*) Ticket marked '1' drawn first, then ticket marked '3'.
(*b*) Ticket marked '3' drawn first, then ticket marked '1'.

TABLE 4.1

Sample size = 2

Sample type	f	Sample mean	Sample type	f	Sample mean	Sample type	f	Sample mean
(1, 2)*	2	1·50	(1, 2, 3)*	6	2·00	(1, 2, 3, 6)*	24	3·00
(1, 3)*	2	2·00	(1, 2, 6)*	6	3·00	(1, 1, 1, 2)	4	1·25
(1, 6)*	2	3·50	(1, 3, 6)*	6	3·33	(1, 1, 1, 3)	4	1·50
(2, 3)*	2	2·50	(2, 3, 6)*	6	3·67	(1, 1, 1, 6)	4	2·25
(2, 6)*	2	4·00	(1, 1, 2)	3	1·33	(2, 2, 2, 1)	4	1·75
(3, 6)*	2	4·50	(1, 1, 3)	3	1·67	(2, 2, 2, 3)	4	2·25
(1, 1)	1	1·00	(1, 1, 6)	3	2·67	(2, 2, 2, 6)	4	3·00
(2, 2)	1	2·00	(1, 2, 2)	3	1·67	(3, 3, 3, 1)	4	2·50
(3, 3)	1	3·00	(1, 3, 3)	3	2·33	(3, 3, 3, 2)	4	2·75
(6, 6)	1	6·00	(1, 6, 6)	3	4·33	(3, 3, 3, 6)	4	3·75
			(2, 3, 3)	3	2·67	(6, 6, 6, 1)	4	4·75
			(2, 6, 6)	3	4·67	(6, 6, 6, 2)	4	5·00
			(2, 2, 3)	3	2·33	(6, 6, 6, 3)	4	5·25
			(3, 6, 6)	3	5·00	(1, 1, 2, 2)	6	1·50
			(2, 2, 6)	3	3·33	(1, 1, 3, 3)	6	2·00
			(3, 3, 6)	3	4·00	(1, 1, 6, 6)	6	3·50
			(1, 1, 1)	1	1·00	(2, 2, 3, 3)	6	2·50
			(2, 2, 2)	1	2·00	(2, 2, 6, 6)	6	4·00
			(3, 3, 3)	1	3·00	(3, 3, 6, 6)	6	4·50
			(6, 6, 6)	1	6·00	(1, 1, 2, 3)	12	1·75
						(1, 1, 2, 6)	12	2·50
						(1, 1, 3, 6)	12	2·75
						(2, 2, 1, 3)	12	2·00
						(2, 2, 1, 6)	12	2·75
						(2, 2, 3, 6)	12	3·25
						(3, 3, 1, 2)	12	2·25
						(3, 3, 1, 6)	12	3·25
						(3, 3, 2, 6)	12	3·50
						(6, 6, 1, 2)	12	3·75
			* Sampling *without* replacement can			(6, 6, 1, 3)	12	4·00
			produce only these sample types.			(6, 6, 2, 3)	12	4·25
						(1, 1, 1, 1)	1	1·00
						(2, 2, 2, 2)	1	2·00
						(3, 3, 3, 3)	1	3·00
						(6, 6, 6, 6)	1	6·00

Similarly a (1, 2, 6) sample can be obtained in 6 ways and a (1, 1) sample in 1 way.

Hence in Table 4.1, *f*, the frequency denotes the number of different sequences which produce a sample of a given type.

It is convenient to regroup these tables as simple frequency tables so that samples having the same mean are grouped together regardless of sample composition. Thus for samples of size 2 the sample types (1, 3) and (2, 2) are grouped together, giving a total frequency $f = 3$ for samples having mean $= 2.00$. Similarly for samples of size 4 the sample type (1, 2, 3, 6), (2, 2, 2, 6) and (3, 3, 3, 3) are grouped together, giving a total frequency $f = 29$ for samples of mean 3·00. The results of these groupings are shown in Table 4.2. In this form these

TABLE 4.2

Sample size = 2 Sample mean (x)	f	Sample size = 3 Sample mean (x)	f	Sample size = 4 Sample mean (x)	f
1·00	1	1·00	1	1·00	1
1·50	2	1·33	3	1·25	4
2·00	3	1·67	6	1·50	10
2·50	2	2·00	7	1·75	16
3·00	1	2·33	6	2·00	19
3·50	2	2·67	6	2·25	20
4·00	2	3·00	7	2·50	22
4·50	2	3·33	9	2·75	28
5·00	0	3·67	6	3·00	29
5·50	0	4·00	3	3·25	24
6·00	1	4·33	3	3·50	18
		4·67	3	3·75	16
		5·00	3	4·00	18
		5·33	0	4·25	12
		5·67	0	4·50	6
		6·00	1	4·75	4
				5·00	4
				5·25	4
				5·50	0
				5·75	0
				6·00	1

tables are exactly comparable to the frequency tables used in Section 2.3 to illustrate the calculation of means and variances. The same calculation procedure applied to these tables gives *means of sample means* and *variances of sample means* because the quantity *x* being manipulated is in each case a sample mean. The results are tabulated in Table 4.3.

T ABLE 4.3

**Means and variances of sample means for all samples
of sizes 2, 3, and 4 (sampling with replacement)**

Sample size	$n = 2$	$n = 3$	$n = 4$
Mean of sample means	3·00	3·00	3·00
Variance of sample means	1·75	1·167	0·875

From Table 4.3 the mean of sample means is seen to equal the population mean (3·00) for each sample size. The variance of sample means can be compared with the theoretical variance of sample means calculated using the formula $\frac{\sigma^2}{n}$.

Sample size	$n = 2$	$n = 3$	$n = 4$
Theoretical variance $\left(\frac{\sigma^2}{n}\right)$	$\frac{3\cdot5}{2} = 1\cdot75$	$\frac{3\cdot5}{3} = 1\cdot167$	$\frac{3\cdot5}{4} = 0\cdot875$

The theoretical calculations agree with those calculated.

These results apply of course to samples obtained with replacement sampling. The corresponding results for samples obtained by sampling without replacement are shown in Table 4.4. The calculations are made on the data marked with

T ABLE 4.4

**Means and variances of sample means for all samples
of sizes 2, 3, and 4 (sampling without replacement)**

Sample size	$n = 2$	$n = 3$	$n = 4$
Mean of sample means	3·00	3·00	3·00
Variance of sample means	1·167	0·389	0

an asterisk in Table 4.1. The theoretical variance is calculated using the formula $\frac{\sigma^2}{n} \cdot \frac{N-n}{N-1}$.

Sample size	$n = 2$	$n = 3$	$n = 4$
Mean of sample means	3·00	3·00	3·00
Variance of sample means	1·167	0·390	0
Theoretical variance	1·167	0·389	0

Again the mean of sample means is shown to equal the population mean and the appropriate variance formula is shown to give results agreeing with calculation.

Numerical demonstrations, as above, do not of course constitute a *proof* of the relationships offered. The reader is asked to accept that what is shown to be true for the specified sizes of sample drawn from a particular population is true for all random samples of this type from all populations satisfying the assumptions.

The final relationship offered, that the distribution of sample means approaches normal as sample size increases, is an expression of what statisticians know as the Central Limit Theorem. Though this is a result of outstanding importance and usefulness in sampling theory formal proof is beyond the scope of this book. The results tabulated in Table 4.2 can be presented in such a way as to make the relationship believable. This is done in Figure 4.1. The frequencies of occurrence

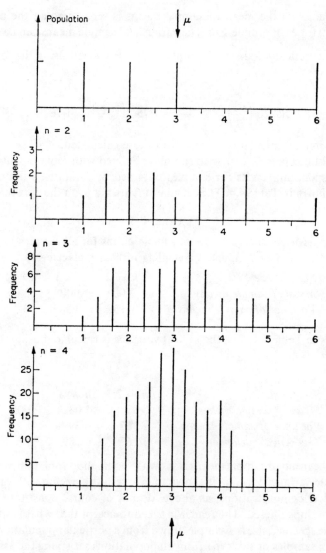

Figure 4.1   Frequency of occurrence of population values and of sample means

of sample means for each sample size are plotted on a common axis together with the population values (which may be thought of as means of samples of size 1). These frequency diagrams show that as the sample size increases the original asymmetric distribution becomes replaced by distributions progressively more symmetrical and apparently single-peaked. The reader is asked to accept that this approach to a symmetrical single peaked distribution continues as sample size increases until a distribution indistinguishable from Normal is reached. Each of these distributions will have a mean equal to the original population mean and a variance equal to the population variance divided by the sample size.

4.3.1 EXAMPLE OF EFFECT OF VARYING SAMPLE SIZE

Coffee is sold in jars of nominal 250 grams content. Jars are filled automatically by a machine which can be pre-set to deliver μ grams, but which will produce jars whose contents will vary approximately normally about this mean of μ and with a standard deviation of 10 grams. If it is known that quality control inspectors take a random sample of 16 jars and condemn the batch being despatched if the mean weight of that sample is less than 245 grams, what machine setting should be adopted to ensure that not more than 1 batch in 1,000 is rejected?

As the population of weights is approximately normally distributed it is reasonable to assume that the means of samples of size 16 will be normally distributed with mean μ and standard deviation $10/\sqrt{16} = 2 \cdot 5$ grams. If the required machine setting is μ, then μ must be the mean of a Normal distribution $N(\mu, 2 \cdot 5)$ such that the tail of the distribution below the value 245 contains $0 \cdot 001$ of the weights. From Table T.1 it can be seen that a value of $z = 3 \cdot 08$ is required, i.e. 245 must be $3 \cdot 08$ standard deviations from μ.

Hence $\dfrac{\mu - 245}{2 \cdot 5} = 3 \cdot 08$

Therefore $\mu = 245 + 7 \cdot 7 = 252 \cdot 7$

Thus the machine must be pre-set to fill jars with $252 \cdot 7$ grams if the specified risk of rejection is not to be exceeded.

4.4 SAMPLING DISTRIBUTIONS

The foregoing discussion of the distribution of sample means is usually entitled the *sampling distribution of means*. Similar terms are used to describe the distribution of any statistical measure whose value is estimated from a sample of data. Thus the terms *sampling distribution of the median* and *sampling distribution of the correlation coefficient* may accordingly be used. In these cases a *sample of data* provides values which are manipulated according to a prescribed formula to obtain an estimate of the value of some desired feature. In all such cases a variant in nomenclature is used which can cause confusion to the statistical beginner, namely the change from *standard deviation* to *standard error*. The term σ^2/n gives the variance of the distribution of sample means. It is natural to call the square root of this variance the standard deviation of the distribution of sample means or more simply the *standard deviation of the mean*. When sampling distributions are being considered this property is referred to as the *standard error*. In this particular case one would refer to the *standard error of the mean*. The introduction of the term 'error' is most easily understood by recognizing that the property reflects the uncertainty associated with sample based estimates. It is emphasized that standard deviation and standard error describe exactly the same property of a distribution. The distinction between them is the distinction between the *distributions* to which they are applied. Standard deviation is a general term; standard error is, by convention, applied to sampling distributions.

4.5 ESTIMATION AND ESTIMATORS

Sampling is carried out to enable estimates to be made of characteristics of the population. Most often the mean and variance of the population are the characteristics whose values are to be estimated. It is apparent that any single sample *may* be unrepresentative of the population even to the extent of being a freak sample. However, if all that is known about the population is information contained in the sample it is impossible to recognize whether or not that sample is a freak. Despite this it is desirable to be able to claim that the estimate made from a given sample is a 'good' estimate and to be able to assess the confidence which can be placed in the value estimated.

These problems come within the province of *estimation theory*. The arguments used involve considering the properties of *estimators* rather than of *estimates*. This distinction is between the formula (or estimator) which is used to convert a set of sample data into a particular value (or estimate). In other words the estimator is the *procedure* used to obtain an estimate. Estimation theory considers the properties of such procedures and when these properties enable an estimator to be described as 'best', then 'best' is also applied to the estimates obtained. Some desirable properties of estimators are discussed briefly in Appendix B.

The essential results are that for practical purposes:

(*a*) The sample mean is the 'best' estimator of the population mean.

(*b*) The sample variance $\times n/(n-1)$ is the 'best' estimator of the population variance (where *n* is, as usual, the sample size).

4.5.1 RELIABILITY OF ESTIMATES

Two distinct points can be made about reliability:

(*a*) Because, for sampling with replacement, the sampling distribution of the mean has a variance of σ^2/n it can immediately be recognized that the precision of any estimate of population mean, defined by the probability of the estimate lying close to the true mean, depends only on population variance, σ^2, and sample size, n and *not* on any relationship between population size and sample size. Essentially this means that if two populations P_1 and P_2 have the same variance but P_1 contains 100 members and P_2 10,000 members, then random samples of size 10, say (with replacement) from each population will give equally precise estimates of the respective population means. This result is *not* obvious because intuitively, but wrongly, one is inclined to suppose that the larger the population the larger the sample should be.

(*b*) Because the distribution of means of all but the smallest samples can be usefully approximated by a Normal distribution with mean μ and standard error $\sigma/\sqrt{n}$ the properties of the Normal distribution can be employed to make statements about the probability of a sample having a mean more than a specified number of standard errors from the true mean μ. It follows from the discussion of the Normal distribution in Chapter 3 that:

1. 67% of sample means will lie between $\mu + \sigma/\sqrt{n}$ and $\mu - \sigma/\sqrt{n}$
2. 95%   ,,   ,, $\mu + 1{\cdot}96\sigma/\sqrt{n}$ and $\mu - 1{\cdot}96\sigma/\sqrt{n}$
3. 99%   ,,   ,, $\mu + 2{\cdot}57\sigma/\sqrt{n}$ and $\mu - 2{\cdot}57\sigma/\sqrt{n}$

It is reasonable to invert these statements in order to give *confidence limits* for any estimate of population mean.

In simplified form the argument would be that if for a sample size n there is a probability of 0·95 of finding the mean within $\mu \pm 1{\cdot}96\ \sigma/\sqrt{n}$, then for any single sample of mean m there should be a probability of 0·95 of finding μ within $m \pm 1{\cdot}96\ \sigma/\sqrt{n}$. (Analogous statements would be made for any desired probability line.) In practice confidence limits are *calculated* according to this argument but the *interpretation* of these limits is not so simple as that suggested above. Essentially the difficulty arises because though one may properly speak of a distribution of sample means one may *not* speak about a distribution of μ, the population mean. The population mean, though unknown, is a single value and thus either falls within the range of the confidence limits or does not and attaching a probability to this is, except in a special sense, artificial. The special sense is that if the process of constructing confidence limits in this way is repeated many times, then 95% of such intervals will contain the population mean μ. (The reader is asked to note that he may encounter an alternative approach to this problem under the title 'fiducial values'. Though in most cases this leads to exactly the same numerical result this is not so for all cases and the logical foundations of the method are different.)

Example. A random sample of 25 orders are examined and the average number of items per order is 28·27 with a variance of 5·28. What is the 95% confidence limit of mean number of items/order?

Estimated population mean $= 28\cdot27$
Estimated population variance $= 5\cdot28 \times 25/24 = 5\cdot50$

Hence estimated standard error of sample means $= \sqrt{\dfrac{5\cdot5}{25}} = \sqrt{0\cdot22} = 0\cdot47$

95% Confidence limits $= 28\cdot27 \pm (1\cdot96 \times 0\cdot47) = 28\cdot27 \pm 0\cdot92$.
$$= 27\cdot35 \text{ to } 29\cdot19.$$

(Strictly the '*t*' distribution should be used here, *see* Chapter 6.)

4.6 SAMPLING OF PROPORTIONS

In Section 4.3 the basic features of sampling theory are demonstrated in terms of sampling a variable which can take values in a range of values, i.e. a continuous variable. The theory developed allows one to estimate the average value of that variable in the population and to assign confidence limits to that estimate. A special case exists when the variable can only take values 0 and 1 and this case is most commonly encountered in the form of sampling to estimate the *proportion* of a population having a particular characteristic.

Suppose the manufacturers of Brand X wish to estimate their share of the market in a particular locality. Subject to certain common-sense assumptions about the nature of the product one way of estimating market share would be to interview a random sample of shoppers and ask whether they buy Brand X or one of the competing brands. A record of these interviews might naturally code purchasers of Brand X as '1' and purchasers of another brand as '0'. If n shoppers are interviewed and r '1's are recorded, then the proportion of Brand X purchasers in the population (and indirectly market share) is estimated as r/n. It is desirable to be able to assign confidence limits to this estimate and results analogous to those in Section 4.5 are available.

4.6.1 THEORY OF SAMPLING PROPORTIONS

The concepts of proportion and probability are for many practical purposes interchangeable. In the market share illustration in the preceding section it might be that of the population of shoppers buying that type of product, 23% buy Brand X and 77% buy some other brand. For any single shopper chosen randomly from this population it is then reasonable to state that the probability of her being a Brand X buyer is 0·23. Accepting this it is now possible to view the taking of a sample of size n as an operation satisfying the conditions for applying the Binomial distribution formula.

In Chapter 3 the Binomial distribution formula was shown to give the probability of r successes in n trials. In the problem considered it is easily seen that a sample of n independent interviews can be considered as a series of n trials and that r shoppers who admit to buying Brand X can be considered as r successes. The probability of success is, as explained, the proportion, p, of Brand X buyers in the population being sampled. Taking this view:

$$P \; (r \; \text{Brand } X \text{ buyers in } n \text{ interviewed}) = {}^nC_r \, p^r \, (1 - p)^{n-r}$$

This enables one to write down directly:

Average number of Brand X buyers in $n = np$
Standard deviation of Brand X buyers in $n = \sqrt{np \, (1 - p)}$

But the *proportion* of Brand X users is r/n. This is simply the Binomial variable divided by the constant n. Hence:

Average proportion of Brand X buyers in $n = np/n = p$

Standard deviation of proportion of Brand X buyers in $n = \dfrac{\sqrt{np \, (1 - p)}}{n} =$

$$\sqrt{\frac{p(1 - p)}{n}}$$

For all possible samples of size n the mean of sample proportions $= p$.
For all possible samples of size n the standard error of sample proportions $=$

$$\sqrt{\frac{p(1 - p)}{n}}$$

These results correspond exactly to the results given in Section 4.3.

In the case of small samples the probability of a sample having a given proportion can be calculated directly using the Binomial distribution. For large samples the Normal approximation to the Binomial distribution can be used and the conditions under which this approximation is adequate are as discussed in Chapter 3.

Example. In a shopper survey 80 out of a total of 400 interviewed are Brand X purchasers. What are the 95% and 99% confidence limits of the estimate of proportion of shoppers buying Brand X?

Estimated Proportion $(p) = 80/400 = 0.20$

Standard Error of estimate $= \sqrt{\dfrac{0.20 \times 0.80}{400}} = 0.02$

Using the normal distribution approximation:

95% Confidence limits $= 0.20 \pm (1.96 \times 0.02) = 0.161$ to 0.239
99% Confidence limits $= 0.20 \pm (2.57 \times 0.02) = 0.149$ to 0.251.

4.7 SUMMARY

Knowledge of the mean and variance of a population from which random samples of size n are drawn prescribes the mean and variance of the distribution of means of these samples. Sampling requires in general that this logic be reversed. A given sample is used to estimate the mean and variance of the population and, using the associated distribution of sample means, statements can then be made about the uncertainty associated with the estimate of the population mean. The importance of this last step can hardly be overemphasized. Quotation of a calculated mean is of little value unless accompanied by a measure of the uncertainty associated with the mean whether expressed as standard error or as confidence limits.

In this chapter extensive reliance is placed on the use of the Normal distribution. Where a population variance is estimated from a sample it is theoretically necessary to use the 't' distribution instead of the Normal distribution. This usage is demonstrated in Chapter 6, where it is seen that though this distinction is important for small samples, for large samples the use of one set of tables rather than the other leads to virtually the same conclusion.

EXERCISES

1. Suppose random samples of sizes (a) 5, (b) 20, and (c) 100 are drawn from a population whose variance $\sigma^2 = 10$ units squared. What will be the standard errors of the corresponding sample means?

2. Suppose the population in (1) is finite and contains 250 items. What then will be the standard errors of random samples of sizes (a) 5, (b) 20, and (c) 100 drawn from this population?

3. Reconsider the coffee jar example in Section 4.3.1 when the following changes in conditions are made:

(a) the standard deviation is 12 grams.
(b) the inspectors base rejection decisions on samples of 64 jars.
(c) both modifications (a) and (b) are made simultaneously.

4. What probabilities can be attached to claims that the following randomly selected samples originate from a normally distributed population of mean 60 units and standard deviation 10 units?

(a) 61·5, 58·5, 62·5, 61·5
(b) 62·1, 61·0, 59·5, 63·5, 62·0, 63·0, 62·0, 58·5, 61·0
(c) 64·0, 60·0, 70·0, 70·0, 66·0, 64·0, 68·0, 69·0, 69·0

(Note the relationship between this question and the type of argument presented in Chapter 6.)

5. (*a*) Calculate the 99% Confidence Limits of the mean in the example in Section 4.5.1.

(*b*) Re-calculate the 95% and 99% Confidence Limits of this example assuming that the mean and variance are obtained from a random sample of 100 orders.

6. Calculate the 95% Confidence Limits for the estimated proportion of Brand X purchasers where:

(*a*) 8 out of a sample of 40 shoppers are Brand X purchasers,

(*b*) 800 out of a sample of 4,000 shoppers are Brand X purchasers.

5. Sampling Processes

5.1 INTRODUCTION

Samples are generally taken as a step in a decision-making process. If the sample information leads to a 'bad' decision a penalty cost of some kind is incurred. The quality and quantity of information obtained from samples can always be improved by spending more money on the sampling operation. If the balance between penalty costs and sampling costs is to have any meaning it is important that any given level of expenditure on sampling should be utilized as efficiently as possible. This involves considering the minimum sample size necessary to give estimates of required precision and exploiting any features of the population which may make particular selection procedures appropriate.

5.2. SAMPLE SIZE AND CONFIDENCE LIMITS

In Chapter 4 the argument is presented by which confidence limits can be calculated within which the true value of the population mean will lie with a specified probability. In that argument the sample size was predetermined. It is possible to invert this argument to calculate the minimum sample size required to produce confidence limits of a predetermined width.

Example. In the coffee jar example (*see* Section 4.3.1) it might be suspected that the filling mechanism was misbehaving with respect to the *average weight* of coffee fed to each jar though the *variability* of weight (measured by the standard deviation of 10 grams) was unaffected. What size of sample would have to be tested so that the associated 95% Confidence Limits will have a width of $\pm$ 4 grams?

Suppose the required sample size is n.

Accepting that sample means will have a distribution $N(\mu, (10/\sqrt{n}))$ it is possible to write down 95% confidence limits for μ

$$= m \pm 1 \cdot 96 \times 10/\sqrt{n} = m \pm 19 \cdot 6/\sqrt{n}, \text{ where } m \text{ is the sample mean.}$$

As these confidence limits must have the form $m \pm 4$ it follows that $4 = 19 \cdot 6/\sqrt{n}$

$$\therefore \sqrt{n} = 4 \cdot 9$$
$$\therefore \quad n = 4 \cdot 9^2 = 24 \cdot 01$$

Therefore a sample size of 25 is necessary to obtain the required precision. (Consideration will convince the reader that 'rounding up' rather than 'rounding off' is necessary.)

Note that the use of the Normal distribution presupposes that the sample size found will be large enough to justify the assumption that sample means will be normally distributed.

5.3 SAMPLE SIZE AND COST

It is always desirable to be able to find the economic balance between sampling costs and bad decision penalty costs. The usual difficulty is in assessing the penalty costs. Two examples follow which illustrate how this balance may be calculated given different assumptions about the nature of penalty costs.

5.3.1 QUADRATIC PENALTY COST FUNCTION

The cost, C_n, of taking a sample of size n can usually be written as $C_n = C_1 + nC_2$ where $C_1 = $ set up cost of the sampling operation incurred regardless of whether or not the sample contains 1 item or 101 items and $C_2 = $ cost of each unit of the sample. The penalty cost, C_p, can reasonably be assumed to be a function of sample size such that the penalty cost *decreases* as the sample size

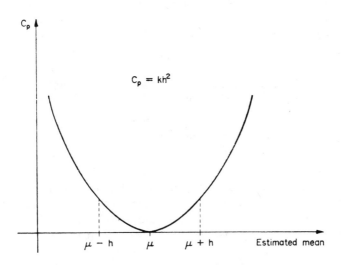

Figure 5.1 Quadratic curve

increases because the larger the sample the more reliable the inferences drawn from it. If it can be assumed that this function is of the *quadratic* form shown in Figure 5.1, then a simple formula relating sample size to total cost can be deduced. ('Quadratic' is an adjective used to indicate that one quantity varies proportionately to the *square* of another. A simple form of such a function is written $y = kx^2$.) The Figure shows a quadratic curve of C_p plotted against the estimated mean. If the estimate coincides with the true mean μ, then $C_p = 0$, but if the estimate deviates from the true mean by an amount h, say, then $C_p = kh^2$.

In the case where a sample has a mean, m_1, then $C_p = k(\mu - m_1)^2$ where k is a proportionality constant expressing units of cost. If one considers all possible samples of a given size n, then the average squared deviation of the sample means from the true mean (i.e., of all terms of the kind $(\mu - m_i)^2$) is of course the variance of the sample means and this is known to be σ^2/n. Thus one can write the average penalty cost associated with a sample of size n as $C_p = k\sigma^2/n$. Combining this with the other costs specified one can write the *total expected cost*, T_c, associated with taking a sample of size n as:

$$T_c = C_1 + nC_2 + \frac{k\sigma^2}{n}$$

It is shown in Appendix C that the value of n which minimizes this total cost is given by:

$$n = \sqrt{\frac{k\sigma^2}{C_2}}$$

The quadratic assumption is very difficult to establish as an *accurate* measure of the way costs vary with error. It is sometimes a useful approximation when it is recognized that costs increase more rapidly than in direct proportion to the magnitude of error. Thus a small overestimate of sales may incur penalty costs only by a reduction in turnover rate. Larger overestimates may involve disproportionately higher costs due to the need to scrap unsold items and to undertake additional advertising and perhaps to close down production units. Similar escalating costs can be attributed to underestimates. In such cases a quadratic cost function *may* be an adequate approximation.

Example. A petrol company is considering embarking on a 'give away' promotional scheme which it is believed will encourage motorists to purchase the company's brand (on the assumption that this does not increase the number of litres bought per motorist). The value of the give-away coupons must be related to average number of litres purchased per motorist and a sample survey is to be mounted to obtain an up-to-date estimate of this quantity. The variance of purchases is believed to be of the order of 5 (litres/day)^2 from previous surveys and the cost of error is believed to be quadratic with constant $k = £100,000$.

Each motorist interviewed involves a cost of 10p ($C_2 = £0{\cdot}1$). What is the optimal number of motorists to be interviewed?

(N.B. $k = £100,000$ implies that an error of $\pm\ 0{\cdot}1$ litres/motorist will cost the company $£100,000 \times (0{\cdot}1)^2 = £1,000$, while an error of $\pm\ 0{\cdot}5$ litres/motorist will cost $£100,000 \times (0{\cdot}5)^2 = £25,000$.)

$$k = £100,000: \sigma^2 = 5; C_2 = 0{\cdot}1$$

$$n = \sqrt{\frac{100,000 \times 5}{0{\cdot}1}} \qquad = \sqrt{5,000,000} \qquad = 2,236$$

5.3.2 SAMPLE SIZE AND A NON-ALGEBRAIC PENALTY COST

The assumption of a quadratic cost function gives a tidy formula but is not readily established in practice. It is frequently possible to calculate the optimal sample size by enumeration of various probabilities in the range of values.

Example. A company manufacturing aircraft structural components supplies them in batches of 50 at a price of £120/item of which £35 is profit. Each batch supplied must contain n extra components supplied free by the manufacturer for destructive testing. The contract requires that the average life of the tested components should be at least 300 h, otherwise the batch is rejected. The manufacturer believes that the test life of the components is normally distributed with mean 330 h and standard deviation 30 h. If a batch is rejected its 'scrap' value is £10/item. How large a test sample should the manufacturer contract to supply?

The manufacturer must balance the cost of supplying free items to the customer against the reducing probability of having a batch rejected as sample size increases. For any test sample of size n the position can be stated.

Manufacturer's expected profit = Profit on 50 items × Probability of 50 items being accepted
− Loss on 50 rejected items × Probability of 50 items being rejected
− Cost of supplying n test items

Profit on 50 items = 50 × £35 = £1,750
Loss on 50 rejected items = 50 × £75 = £3,750
Cost of supplying n test items = n × £85

The probability of a batch being rejected on the basis of a sample of size n is the probability of the average test life of n items being less than 300 h. The lives of individual items are normally distributed so that sample means will be normally distributed regardless of sample size. The following results can be quickly calculated using tables of Normal distribution.

Sample size (n)	S.E. of sample mean $\sigma/\sqrt{n}$	z $\dfrac{(330-300)}{S.E.}$	Prob. mean $\geqslant 300$	Prob. mean < 300
2	21·22	1·41	0·9207	0·0793
3	17·32	1·73	0·9582	0·0418
4	15·00	2·00	0·9772	0·0228
5	13·42	2·24	0·9875	0·0125
6	12·25	2·45	0·9929	0·0071

Applying these probabilities to the profit and loss figures and taking the cost of free test samples into account the picture of manufacturer's *expected* net profit for different sample sizes becomes:

Sample size	Expected profit	Expected loss	Cost of test sample	Expected net profit
2	£1,611·225	£297·375	£170	£1,143·85
3	£1,676·850	£156·750	£255	£1,265·10
*4	£1,710·100	£85·500	£340	£1,284·60*
5	£1,728·125	£46·875	£425	£1,256·25
6	£1,737·575	£26·625	£510	£1,200·95

It can be seen from the right-hand column that the manufacturer's expected total profit is maximized if he agrees to supply four free test pieces for destructive testing.

5.4 SAMPLING SCHEMES

Sample size is one of two features of sampling processes which are variable. The other is the manner of selection of the sample or *sampling scheme* adopted. Essentially this involves exploiting knowledge of special features of the population. A variety of such schemes are available but only two will be considered here, one in depth. (*See* Refs. 1 and 2 for further methods.)

5.4.1 SYSTEMATIC SAMPLING

Random sampling has been discussed in terms which imply the existence of a population whose members are all equally conveniently accessible at any time. Thus if a random *number* sequence is 593, 164, 884 . . . it is implied that there is no inconvenience in taking firstly item 593, then the lower numbered item 164, and then the even higher numbered item 884, etc. With some populations this is either impracticable or extremely inconvenient. Examples of such populations are (*a*) items coming off a continuous production line, (*b*) households situated on a long street, and (*c*) items recorded in a file or on ledger pages. All such populations have in some sense a sequential existence. It is naturally convenient having selected a particular member of the population, that the next number selected be 'later' in that sequence. In such cases *systematic sampling* may be employed. In the production line context a typical systematic sampling procedure would be:

(*a*) Decide what proportion of the population is to be sampled. This is a decision of the kind: '1 in 10 members of the population will be sampled' and is made perhaps on the basis that a sample size of *n* is required from a given shift's production and a shift is known to produce approximately *N* items, the ratio n/N being 1/10.

(*b*) Using random number tables select a number between 01 and 10, say 06.

(*c*) The items forming the sample will then be 06, 16, 26, 36, 46, etc. (Had a sample of 1/100 been required the items selected, given the same starting number would have been 006, 106, 206, 306, etc.)

(*d*) Provided that the sequence of items produced can be considered as randomly ordered with regard to the characteristic which is being sampled the data collected can be treated as a random sample and the results of Chapter 4 applied.

N.B. One particular type of ordering must be guarded against—periodic ordering. Examples of this kind of ordering are easily imagined.

Example (1). In the production line context it might be that the production line being sampled is 'fed' by 3 input lines *A*, *B*, and *C*. The effect is that the 1st, 4th, 7th, etc., item on the line being sampled originates from feed line *A*; the 2nd, 5th, 8th, etc., from *B* and the 3rd, 6th, 9th, etc., from line *C*. If the systematic sampling ratio is 1/3 or 1/6 or 1/9, etc., then it will be recognized that all items in the sample will originate from the same feed line, the particular line being identified by the first random number selected.

Example (2). In the household context it is usual for odd and even numbered houses to be on opposite sides of the street. It is not uncommon for houses on one side of a street to be systematically different in character from those on the other side. If a systematic sampling procedure has an 'even' sampling ratio, say 1/6 or 1/10 or 1/24, then if the starting number selected is odd all subsequent numbers included in the sample will be odd, whilst if the starting number is even all subsequent numbers will be even and a consequent possible bias is introduced. If the sampling ratio is 'odd', say 1/3, 1/7, 1/21, this difficulty does not arise.

In brief, the principal hazard in systematic sampling is the possible existence of some periodic effect in the population which coincides with a multiple of the sampling ratio. If this can be guarded against the data collected by this process can be treated as a simple random sample. The advantage of the method is that sample selection can be carried out in a progressive sequential manner matching the 'structure' of the population.

5.4.2 STRATIFIED RANDOM SAMPLING

In Section 5.4. the phrase 'exploiting known features of the population' is used. Generally such knowledge is qualitative or at best only roughly quantitative. One type of knowledge which frequently exists is knowledge of ways in which members of the population may be classified. In simplest form a classification of human beings into adult male, adult female, and children would be a common-sense procedure. In a survey of brands of pipe tobacco preferred a sample of 500 interviews would harvest much more useful information if the interviews were restricted to adult males than to human beings as an unclassified group. Similarly a survey of cigarette brands preferred would be better concentrated on adult males and adult females (though whether or not children should be completely ignored in such a survey might depend upon the age definition of 'children' adopted!). Such a classification procedure is called *stratifying* a population and implies simply that a population is divided into classes or *strata* in such a way that each member of the population belongs to one and only one stratum. If the result of such a classification is to produce strata in each of which the members are more like *each other* with respect to the characteristic being sampled than they are like members of other strata, then sampling economies can be effected by treating each stratum as a sub-population. Examples of strata which are used for particular purposes are:

(i) Socio-economic classes form strata within which the members are more alike with respect to spending power, spending habits, and educational background.

(ii) Classification of human beings by age and sex divides them into strata whose members are more alike with respect to life expectancy. Such a stratification would be relevant to the interests of life insurance companies.

(iii) Stratification of car owners by type of residential area (i.e. by urban, suburban or rural area) is useful to motor insurance companies in assessing risk.

(iv) Stratification of retail outlets by size and type (i.e. supermarket to corner shop) is useful in sampling stock turnover of items and is so used in retail audits.

5.4.3. EFFECTS OF STRATIFICATION

Consider the population of equal proportions of units valued 1, 2, 3, and 6 used in Chapter 4 to illustrate the properties of sample means. It will be recalled that the population mean, $\mu = 3 \cdot 0$ and the population variance, $\sigma^2 = 3 \cdot 5$. In Chapter 4 this population is represented in the simple physical form of 4 tickets in a bowl numbered with the values 1, 2, 3, and 6. The contents of the bowl are assumed unknown to the sampler who is trying to estimate the mean of the population on the basis of the means of samples drawn. In a modified version of this 'experiment' the reader is asked to suppose that the two lower value tickets 1 and 2 are placed in one bowl (*A*) and the two higher value tickets 3 and 6 in another bowl (*B*). Sampling with replacement from both bowls is allowed. For samples of size 2 one ticket is drawn from bowl *A* and 1 from bowl *B*. For samples of size 4 two tickets are drawn from bowl *A* and two from bowl *B*. Table 5.1

TABLE 5.1

Sample size $n = 2$			Sample size $n = 4$		
Sample type	*Probability*	*Sample mean*	*Sample type*	*Probability*	*Sample mean*
(1, 3)	0·25	2·00	(11, 33)	0·063	2·00
(1, 6)	0·25	3·50	(11, 66)	0·063	3·50
(2, 3)	0·25	2·50	(11, 36)	0·125	2·75
(2, 6)	0·25	4·00	(22, 33)	0·063	2·50
	Mean of means =	3·00	(22, 66)	0·063	4·00
			(22, 36)	0·125	3·25
			(12, 33)	0·125	2·25
			(12, 66)	0·125	3·75
			(12, 36)	0·250	3·00
				Mean of means =	3·00

$$n = 2 \quad \text{Mean of means} = 3·00$$
$$\text{Variance of means} = 0·6250$$
$$n = 4 \quad \text{Mean of means} = 3·00$$
$$\text{Variance of means} = 0·3125$$

records all possible samples of sizes 2 and 4, together with frequency of occurrence in exactly the same way as results are recorded in Table 4.2.

The mean of sample means still equals the population mean but the variance of sample means is reduced. That this should be so is easily recognized. If the bowls are labelled Low and High as a mnemonic convenience, then:

for $n = 2$ Low, High samples are possible but *not* Low, Low nor High, High samples.

for $n = 4$ Low, Low, High, High samples are possible but not Low, Low, Low, Low, nor High, High, High, High, nor Low, Low, Low, High nor Low, High, High, High.

The effect is to produce samples whose means lie closer to the true mean and to suppress the extremely high and extremely low valued samples thus reducing the variance of sample means. (Ex. $n = 4$, 1111 and 6666 are amongst the impossible samples.)

5.4.4 PARTITIONING OF A SAMPLE AMONGST DIFFERENT STRATA

It is natural to ask at this point whether in the selection of a random sample of size 4 the selection of 2 items from the 'low' group and 2 items from the 'high' group was the *best* way of partitioning the 4 items between the two strata. It would for instance be reasonable to consider taking a sample of 3 from one group and a sample of 1 from the other. Before answering this question it is useful to be clear about the source of advantage in stratification.

Essentially the idea of stratification is to produce groups containing members as nearly similar as possible. If a population of 120 items contained 40 items of value 1 unit, 40 items of value 2 units, and 40 items of value 3 units it would be natural to divide the population into 3 strata. One stratum would contain items of value 1, one stratum items of value 2, and one stratum items of value 3. In such case a sample of size 3 would, if equipartitioned amongst all three strata, provide an accurate estimate of the population mean because every such sample would be of type (1, 2, 3). This is an example of an ideal stratification situation only realized in most exceptional circumstances. It is ideal because:

(*a*) the strata are equal in size;
(*b*) each stratum has zero variance (because its members are identical).

Consider what happens if the actual circumstances differ from this ideal.

5.4.5 STRATA OF UNEQUAL SIZE AND ZERO VARIANCE

Suppose the population of 120 items postulated in the previous section had consisted of 60 items valued 1 unit, 40 items valued 2 units, and 20 items valued 3 units. The population again divides naturally into 3 strata, within each stratum the items are again identical but the number of items in each stratum is different.

If a sample of 3 items is equipartitioned amongst the strata its composition is again of type (1, 2, 3). The sample mean is hence 2·0 but the actual population mean is:

$$\frac{(60 \times 1) + (40 \times 2) + (20 \times 3)}{120} = \frac{200}{120} = 1 \cdot 67$$

The sample mean over-estimates (in this case) the population mean because in the sample equal *weight* is given to each value, whilst in the population unequal proportions of the different values exist. It can easily be seen that if the relative sizes of the different strata are known this weighting can be applied to the sample values. Thus:

$$\text{Weight to be attached to value } 1 = \frac{60}{120} = 0 \cdot 500$$

Weight to be attached to value 2 $= \dfrac{40}{120} = 0{\cdot}333$

Weight to be attached to value 3 $= \dfrac{20}{120} = 0{\cdot}167$

$\therefore$ Weighted sample mean $= (0{\cdot}500 \times 1) + (0{\cdot}333 \times 2) + (0{\cdot}167 \times 3)$
$$= (0{\cdot}500) + (0{\cdot}666) + (0{\cdot}501)$$
$$= 1{\cdot}67$$

This is a general procedure and the weighting factor W_i to be applied to the sample value from the ith stratum is given by

$$W_i = N_i/N$$

where $N_i =$ number of items in the ith stratum
$N =$ number in the population

An equivalent procedure is used when considering how a sample might be partitioned amongst the strata if equi-partitioning is abandoned. Suppose that a sample of size 6 is to be partitioned amongst these 3 strata. If the same weighting factors are applied to the sample *size* instead of the sample values and the sub-sample sizes so obtained drawn from the appropriate strata the result is as shown.

Sample size $= 6$. Weighting factors: Stratum 1 $= 0{\cdot}500$
Stratum 2 $= 0{\cdot}333$
Stratum 3 $= 0{\cdot}167$

Size of sub-sample to be drawn from statum 1 $= 6 \times 0{\cdot}500 = 3$
" ". " " 2 $= 6 \times 0{\cdot}333 = 2$
" " " " 3 $= 6 \times 0{\cdot}167 = 1$
Hence sample type $= (1, 1, 1, 2, 2, 3)$
Sample mean $= \dfrac{10}{6} = 1{\cdot}67$

This again is a general procedure and the rule is that the number of sample items, n_i, to be drawn from the ith stratum is given by:

$$n_i = \frac{n \times N_i}{N}$$

where n and N are the sample and population sizes respectively and N_i is the number of items in the ith stratum.

Note that even if a population is infinite in size (i.e., $N = \infty$) it is still possible to know that the ith stratum contains $x\%$ of the population and in such a case $x/100$ is the weighting factor (N_i/N), e.g., a continuous production line in a car factory may represent a conceptually infinite population of cars but it is still possible to say that 30% of these cars are white in colour.

5.4.6 STRATA OF NON-ZERO AND UNEQUAL VARIANCE

Suppose a population can be divided into 2 equal-sized strata, one having a variance of 9 units and the other a variance of 36 units. A sample of size 6 is to be partitioned between these strata. How 'best' should this be done?

The purpose of sampling is to obtain an estimate of the population mean. Following on the discussion of estimation given in Chapter 4 it is reasonable to seek a means of partitioning the sample so that the standard error of the estimate of the population mean is as small as possible. If simple random sampling is assumed for each stratum, then the standard error of the estimate of that stratum's mean is obtained by dividing the stratum standard deviation by the square root of the sub-sample size taken from that stratum (remember $\sigma/\sqrt{n}$). The population mean is estimated by the sum of the sub-sample means because the strata are of equal size.

The standard error of this estimate is obtained by adding the squares of the standard error of the individual strata estimates and taking the square root of that sum. (Remember *variances* may be added but *not standard deviations*.) Applying this procedure to the values given above leads to the results shown in Table 5.2.

TABLE 5.2

Effect of different partitioning schemes on Standard Error of Estimates of Population mean

Total sample size = 6

Sample size partition	Stratum A	Stratum B	Standard Error of Estimate of Population mean
	1	5	$\sqrt{\left(\dfrac{9}{1}+\dfrac{36}{5}\right)} = \sqrt{16 \cdot 2} = 4 \cdot 03$
	2	4	$\sqrt{\left(\dfrac{9}{2}+\dfrac{36}{4}\right)} = \sqrt{13 \cdot 5} = 3 \cdot 67$
	3	3	$\sqrt{\left(\dfrac{9}{3}+\dfrac{36}{3}\right)} = \sqrt{15 \cdot 0} = 3 \cdot 87$
	4	2	$\sqrt{\left(\dfrac{9}{4}+\dfrac{36}{2}\right)} = \sqrt{20 \cdot 25} = 4 \cdot 50$
	5	1	$\sqrt{\left(\dfrac{9}{5}+\dfrac{36}{1}\right)} = \sqrt{37 \cdot 80} = 6 \cdot 15$

It is easily seen that the minimum standard error of estimate is obtained when the sample is partitioned so that 2 items are sampled from stratum *A* and 4 from stratum *B*. If one denotes the stratum variances as σ_A^2 and σ_B^2, where $\sigma_A^2 = 9$ and $\sigma_B^2 = 36$, it will be recognized that the sample is partitioned using a 'weighting factor'

$\dfrac{\sigma_A}{\sigma_A + \sigma_B}$ and thus obtaining $n_A = \dfrac{\sigma_A}{\sigma_A + \sigma_B} \times n$

This is a general rule and for more than 2 strata the sub-sample size for the ith stratum is given by:

$$n_i = \frac{\sigma_i}{\sigma_1 + \sigma_2 + \ldots \sigma_n} \times n$$

5.4.7 GENERAL CASE. STRATA OF UNEQUAL SIZE AND WITH UNEQUAL VARIANCE

In the general case both the features discussed in Sections 5.4.5 and 5.4.6 occur together. Strata sizes and strata variances are different. In such a case if the sample size is fixed at n items and each of the strata A, B, $C \ldots$ has size and variance N_A, $\sigma_A{}^2$; N_B, $\sigma_B{}^2$; N_C, $\sigma_C{}^2$; $\ldots$ then the number of sample items to be drawn from stratum A is:

$$n_A = n \times \left(\frac{N_A \, \sigma_A}{N_A \sigma_A + N_B \sigma_B + N_C \sigma_C + \ldots} \right)$$

The result for the simplest case of two strata is derived in Appendix D. It will be recognized that the formula above reduces to proportioning with respect to strata *sizes* when the strata standard deviations are equal and proportioning with respect to strata *standard deviations* when the strata sizes are equal.

Example. A survey is planned to estimate the market for tinned soups in a particular region. It is believed that the population of retail outlets can be classified as supermarkets, multiples, large independents, and small independents, of which there are 10, 35, 60, and 200 respectively. The standard deviation of average weekly sales in each stratum is estimated to be 30, 70, 90, and 25 respectively. How should a sample of 50 shops be apportioned amongst these different types of shop?

Using the formula given above:

No. of supermarkets in sample $= \dfrac{50 \times (10 \times 30)}{(10 \times 30) + (35 \times 70) + (60 \times 90) +}$
$\qquad\qquad\qquad\qquad\qquad\qquad\qquad\qquad\qquad (200 \times 25)$

$\qquad\qquad = \dfrac{15,000}{300 + 2,450 + 5,400 + 5,000}$

$\qquad\qquad = \dfrac{15,000}{13,150} = 1\cdot14$

No. of multiples in sample $\quad = \dfrac{50 \times (35 \times 70)}{13,150} = \dfrac{122,500}{13,150}$

$\qquad\qquad = 9\cdot32$

No. of large independents in sample

$$= \frac{50 \times 60 \times 90}{13,150} = \frac{270,000}{13,150}$$
$$= 20 \cdot 53$$

No. of small independents in sample

$$= \frac{50 \times 25 \times 200}{13,150} = \frac{250,000}{13,150}$$
$$= 19 \cdot 01$$

These values must be rounded off to the nearest whole number giving:

Supermarkets	*Multiples*	*Large independents*	*Small independents*
1	9	21	19

5.4.8 ADVANTAGES OF STRATIFIED SAMPLING

Example. Suppose a population consists of equal numbers of items valued 1, 2, 3, 4, 5, 25, 35, and 45. Calculation shows $\mu = 15$; $\sigma^2 = 266 \cdot 25$.

The population divides naturally into two strata:

Stratum A (1, 2, 3, 4, 5) $\mu_A = 3$ $\sigma_A^2 = 2 \cdot 00$
Stratum B (25, 35, 45) $\mu_B = 35$ $\sigma_B^2 = 66 \cdot 67$

If a simple random sample of size 50 is drawn from the unstratified population its variance $= \sigma^2/n = 266 \cdot 25/50 = 5 \cdot 325$ and hence the standard error of the estimate $= 2 \cdot 31$.

If the sample of 50 is optimally partitioned between strata A and B:

$$n_A = 50 \times \frac{5 \times 1 \cdot 414}{(5 \times 1 \cdot 414) + (3 \times 8 \cdot 17)} = \frac{353 \cdot 5}{7 \cdot 07 + 24 \cdot 51} = \frac{353 \cdot 5}{31 \cdot 58}$$
$$= 11 \cdot 2$$

$$n_B = 50 \times \frac{24 \cdot 51}{31 \cdot 58} = \frac{1,225 \cdot 5}{31 \cdot 58} = 38 \cdot 8$$

Rounding these values off to 11 and 39 the variance of the resulting estimate of the mean is (from the formula in Appendix D):

$$\left(\frac{5}{8}\right)^2 \times \frac{2}{11} + \left(\frac{3}{8}\right)^2 \times \frac{66 \cdot 67}{39}$$
$$= \left(\frac{25}{64} \times \frac{2}{11}\right) + \left(\frac{9}{64} \times \frac{66 \cdot 67}{39}\right) = 0 \cdot 071 + 0 \cdot 244 = 0 \cdot 315$$

Hence standard error of estimate $= 0 \cdot 56$.

It should be understood that even rough and ready estimates of the relative sizes and variances of strata can be utilized in establishing the sizes of the sample components to be drawn from each stratum. In practice an approximation widely used is that of taking the *range* of values in each stratum and using this in place of the standard deviation when calculating the way in which a sample should be partitioned. Information of this kind is usually obtained by either a pilot sampling operation or in the course of the main study. In the latter case a population may be divided into 4 strata (say) and the intention be to take a total sample of 100 items from the population. If initially 6 items (say) are drawn from each stratum and the relative variability of the strata is determined from these 24 items the remaining 76 items may be partitioned in such a way as to bring the strata sub-samples as near as possible to the optimal sizes.

5.5 MISCELLANEOUS SAMPLING PROCEDURES

A variety of sampling procedures are available which are appropriate to particular situations. As exploitation of population characteristics becomes more ingenious so does the mathematical explanation required become more complex. The following brief paragraphs are intended to introduce some of the more commonly occurring sampling terms. The reader wishing to extend his knowledge of these should consult Refs. 1 and 2 where relatively non-mathematical descriptions of a variety of procedures will be found.

5.5.1 CLUSTER SAMPLING

In *Cluster Sampling* the items sampled are groups or clusters of what might be considered the natural items to sample. Thus instead of sampling individuals in a particular area one might sample households. This situation could arise either because a complete sampling frame defined in terms of natural units (in this case individuals) did not exist and was too costly to construct or because the economics of sampling (in this case, travelling costs) dictated a more cost conscious approach. An advantage of this type of sample is that it allows 'before' and 'after' comparisons to be made by follow-up interviews. It is however important to recognize that only the clusters are randomly selected not the individuals constituting them.

5.5.2 QUOTA SAMPLING

This involves synthesizing a sample having the same structure as the population as a whole in terms of characteristics believed relevant to the opinion being sought. Suppose a market survey is being undertaken to determine the relative attractiveness of different features in a women's magazine. It might be believed that age and marital status are the important factors in determining a woman's

attitude to various types of magazine article. If it is known that the age/marital
status of the female population is distributed as follows:

	Age	16–20	21–30	31–40	41–50 ...
Marital	Single	6%	4%	4%	2%
Status	Married	4%	9%	9%	11%
	Divorced	2%	3%	2%	1%
	Widowed	1%	2%	5%	6%
		13%	18%	20%	20% (Hypothetical!)

then a sample would be taken which contained those percentages of each type of
age/marital status female. In practice this might be achieved by a process in
which as each interview took place the views were recorded only if membership
of that particular category in the sample had not yet been filled. Thus if a
sample requires 4 under 20 divorcees and these have been obtained a subsequent interviewee may turn out, *during the course of the inverview*, to constitute
a 5th member of that category. In such a case the interview could either be
brought to a premature stop or continued but the result discarded from the
sample. In general, quota sampling may be described as stratified sampling with
a more or less non-random selection of units within strata. For this reason,
sampling error formulae cannot be applied with confidence to the results of
quota samples.

In most quota sampling a number of cross-classifications are required to
reproduce the population. Representative testing panels are often constructed
on this basis, such as the supposedly representative studio audiences assembled
for some 'serious' television discussion programmes. Quota sampling should not
be confused with the term 'Judgment Sampling'. This usually applies to some
group which is *believed*, often without good cause, to represent the population
in miniature. Examples of this kind crop up in Parliamentary election discussions
when it is sometimes claimed that the winning party can be predicted from the
candidate returned by a particular constituency. A rather better-founded
example of this kind of sampling was demonstrated in the television coverage of
the 1970 General Election when the constituency of Gravesend was selected as
typical of the electorate as a whole and a poll carried out after voting but before
the declaration of any results was used to estimate the 'swing' from Labour to
Conservative over the whole country with considerable success. In this case it
seems likely that belief in the representative nature of the constituency was based
on research rather than simple faith or mythology.

5.6 SUMMARY

Value for money is desirable in every business activity and sampling is no
exception to this dictum. Sample results are information and this information is

only obtained at a price. To decide whether or not the price is reasonable involves consideration of the precision of the estimate which is made from the sample and this can be related to the risks which one is prepared to run in consequent decision-making. If these risks can be 'costed' it is often possible to establish the optimal sample size as a 'trade off' between the cost of sampling and the cost of a bad decision. In any case it may be possible to ensure a better return for money spent on sampling by adopting a sampling scheme which incorporates 'knowledge', even of a rough and ready kind, about the properties of the population being sampled.

EXERCISES

1. For the example in Section 5.2 what sample size would be necessary to ensure that the 95% Confidence Limits were no wider than 2 grams?

2. Estimate the optimal sample size in the example in Section 5.3.1 where:

(a) the estimated variance of purchases is 10 (litres/day)^2;
(b) the cost of interviewing is 20p/interview;
(c) where both the above changes apply simultaneously.

3. What sample sizes are necessary so that the 95% Confidence Limits of the estimated proportion of Brand X users is no more than ± 0.01 if the proportion is believed to be:

(a) of the order of 0.20;
(b) of the order of 0.10;
(c) of the order of 0.50.

What general conclusion do you draw from your results?

4. For the example in Section 5.4.7 show how the sample allocation is changed if the standard deviations of sales in different kinds of shop are 70, 90, 30, and 25 respectively.

6. *Significance Testing*

6.1 INTRODUCTION

Suppose Sloppo Breakfast Cereal Ltd commissions an annual market survey. The most recent results are:

Estimated market share	*95% confidence limits*
Last year 20%	17–23%
This year 19%	17–21%

Is the Sloppo management justified in taking action on the basis of market share having dropped? The single figure estimates differ but the uncertainty in these estimates (expressed as confidence limits) is sufficiently large to raise the suspicion that the difference may just be a sampling effect. Sloppo management is entitled to expect some clarification of their position from their marketing advisers. Significance testing enables these two results to be compared and a statement made to the effect that the difference of 1% should either be heeded or neglected and that the probability of that decision being wrong is less than $x\%$.

This is a specific example of a significance testing application. A variety of significance tests are available and are appropriate to different situations. Some of the more common tests will be explained and illustrated in this chapter, these and others are described in Langley [2]. The reader is particularly requested to note that the *logical framework* in which they are used is virtually the same regardless of the details of the test.

6.2 LOGICAL FRAMEWORK

The logic of significance testing can be illustrated by the following trivial example.

Suppose one is asked to decide whether a coin is 'double-headed' or 'normal' on the basis of the reported results of a sequence of tosses. A rational, and intuitively reasonable, procedure would be as follows:

(*a*) Adopt the hypothesis that the coin is normal (because the vast majority of coins are normal). Such a hypothesis will be referred to as a *null hypothesis* and is so called because it denies the existence of any 'special' effect.

114

(*b*) Consider the results reported in the light of this hypothesis, and, if they appear too improbable reject the hypothesis. Otherwise accept it.

In the situation stipulated the probability of any sequence of results can readily be calculated. Suppose a sequence of heads is reported. If the normal coin hypothesis is true:

Probability of 1 head $= 0.5$, i.e. $(\frac{1}{2})$
Probability of 2 heads $= 0.25$, i.e. $(\frac{1}{2})^2$
Probability of 3 heads $= 0.125$, i.e. $(\frac{1}{2})^3$
Probability of 4 heads $= 0.0625$, i.e. $(\frac{1}{2})^4$

Continuing in this way a level of probability is reached which is so low as to make continued adherence to the initial hypothesis unacceptable. This level of low probability is the *significance level* at which one concludes that the results obtained are *significantly different* from those which would have been expected if the hypothesis were true.

This is the basic logic of significance testing. Several points should be noted here.

(1) The choice of significance level is *subjective*.

(2) A test which leads to the rejection of a hypothesis does not *prove* that the hypothesis is *false* but *demonstrates* that it is improbable. This point can be understood most readily by realizing that because the test involves calculating the probability of occurrence of the observed results if the hypothesis is true, then it must be *possible* for the rejected hypothesis to be true.

(3) Choosing a very low probability level as the significance level reduces the probability of *wrongly rejecting a true hypothesis*, because this means that only the most 'outlandish' results will lead to rejection of the hypothesis. This incorrect decision is referred to as a *Type I error*. The converse error of accepting a false hypothesis is termed a *Type II error*.

This balance of advantage is determined subjectively and usually relies on considerations as to which type of error incurs the more serious penalty.

6.3 DEGREES OF FREEDOM

The probabilities required in significance testing are usually obtained by referring to tables of critical values compiled for the particular test being used. A selection of such tables are given starting on page 235. In referring to these tables relevant rows are selected by using 'indicators' of the kind $n - 1$ and $n - 2$, where n is the number of observations being used in the calculation. These 'indicators' are referred to as the number of *degrees of freedom* in the calculation. The mathematical reasoning behind this concept is too elaborate for this book, but the need for the concept can be illustrated in a very primitive way. To calculate the variance of a set of n numbers a necessary preliminary is the use of the same n numbers to calculate the mean. The deviations of the individual values from this mean are in a sense constrained. Suppose the values are 1, 3, 6, and 10. The mean of these values is 5. The deviations from this mean are -4, -2, 1 and 5. However because the mean is calculated from the set of values the deviations must sum to zero. This being so when *three* of these deviations are known the *fourth* value can be deduced immediately. In this sense, though one starts with four values only three independent deviations result. In the general case n values reduce to $n - 1$ independent deviations in the variance calculations. This is the appropriate number of degrees of freedom. Similar allowances have to be made in all sample-based tests and though the relevant expression will be given in each case it may be convenient for the reader to consider each case as requiring the number of observations to be reduced by the number of necessary prior values or parameters calculated from these observations. An equivalent *aide-mémoire* useful for most tests (though not the chi-squared test) is that the number of degrees of freedom is the divisor in the estimation of the population variance from the sample sum of squared deviations.

6.4 VARIANCE RATIO TEST ('*F*' TEST)

Two sets of sample values are being compared. Sample 1 contains n_1 values with variance s_1^2; sample 2 contains n_2 values with variance s_2^2. Are these variances significantly different?

Procedure:

(1) Adopt null hypothesis that both samples are random samples from a common population (*see* Note A below) and that the apparent difference between the sample variances is due to chance sampling effects.

(2) Decide on a significance level, i.e. a level of disbelief.

(3) Estimate the population variance from each sample separately. These estimates, E_1 and E_2, are:

$$E_1 = \frac{n_1 s_1^2}{(n_1 - 1)} \text{ and } E_2 = \frac{n_2 s_2^2}{(n_2 - 1)}$$

The ratio of these two values is calculated so that it is greater than 1·0. Thus if $E_1 > E_2$ calculate E_1/E_2, but if $E_2 > E_1$ calculate E_2/E_1.

Critical values of this ratio are tabulated for different significance levels (*see* Tables T.4a and T.4b). The *columns* of each table being indexed according to the number of degrees of freedom of the estimated variance above the line in the ratio and the *rows* being indexed by the number of degrees of freedom of the estimated variance below the line in the ratio. (These degrees of freedom are given by $n_1 - 1$ and $n_2 - 1$.)

(4) Select a significance level of $x\%$. Look up the tabulated value corresponding to $\frac{1}{2}x\%$. If the calculated value is greater than the tabulated values the two variances have a probability of less than $x\%$ of originating from the same distribution. Then the null hypotheses is rejected. Otherwise the null hypothesis is retained.

Note A. The tables used in this test are constructed on the basis of the common population being *normally* distributed.

Though it is not usually possible to confirm that this condition is in fact satisfied the reader should be aware that this assumption is implicit in the use of the test.

Example. Two automatic bottling lines have been installed. Though set to fill the same volume a sample of 61 bottles from line A has a variance of 0·0180 units squared whilst a sample of 25 bottles from line B has a variance of 0·0396 units squared. On this evidence should one expect the variability of content from line B to be systematically greater than from line A?

Null hypothesis: Both samples can be considered as coming from a common normally distributed population.

Significance level: 0·02 (or 2%)

Calculation:

$$\text{Estimate of population variance (line } A \text{ sample)} = \frac{0·0180 \times 61}{60} = 0·0183$$

$$\text{Estimate of population variance (line } B \text{ sample)} = \frac{0·0396 \times 25}{24} = 0·0413$$

$$\text{Variance ratio} \quad = \frac{0·0413}{0·0183} = 2·26$$

From the 1% variance ratio table the value in the column labelled 24 degrees of freedom and the row labelled 60 degrees of freedom is 2·12.

Because $2·26 > 2·12$ the difference *is* significant at the 2% significance level and the null hypothesis is abandoned.

N.B. The probability of this conclusion being wrong is less than 2% and hence action based on the conclusion that the two production lines tend to have different variabilities in content will only have that risk of being wrong.

6.5 USE OF THE NORMAL DISTRIBUTION

The central feature of the 'mechanism' of a significance test is the means of calculating probabilities given an initial hypothesis. In many cases the Normal distribution provides such a means. Re-scrutiny of the examples used to illustrate the role of the normal distribution in sampling theory will show that these can be related to the significance testing logic specified.

Example. Your company markets car tyres. Their lives are normally distributed with a mean of 40,000 kilometres and standard deviation of 3,000 km. A change in the production process is believed to result in a better product. A test sample of 64 'new' tyres have a mean life of 41,200 km. Can you conclude that the new product is significantly better than the current one?

Null hypothesis: There is no 'real' difference, i.e. the sample results only differ by chance from the current product performance.
Significance level: 0·01 (or 1%)
Calculation: If the hypothesis is true means of samples of size 64 drawn from the current product should be normally distributed with mean 40,000 km and SE $3,000/\sqrt{64} = 375$ km.

From tables of area under the normal curve the probability of a value 3·2 (i.e. 1,200/375) standard errors above the mean is $1·000 - 0·9993 = 0·0007$.

0·0007 is less than 1% and hence the null hypothesis may be rejected.
Conclusion: On this evidence it may be concluded that the new product is superior to the current one and the probability of this conclusion being wrong is less than 1%.

6.6 THE 't' DISTRIBUTION (STUDENT'S 't')

This distribution is used in place of the normal distribution in situations where small samples of data are available. Essentially the 't' distribution should be used in tests of means where the *population variance must be estimated from sample data*. In practice it will be found that where population variance estimates are based on large samples ($n \geqslant 40$) little difference, if any, in conclusion will result whichever distribution is used. For small samples the differences may be substantial. The most common application of the 't' distribution is in testing whether or not two sample means are significantly different, but the 't' distribution should also be used in calculating confidence limits for means of small samples.

6.6.1 DIFFERENCE BETWEEN TWO MEANS—PAIRED COMPARISONS

In some situations two samples of results are obtained in which *each member* of one sample may be compared directly with a *corresponding member* of the other sample. Examples are:

(*a*) Comparison of sales in a set of shops before and after a sales campaign.
(*b*) Comparison of salesmen's expenses claims before and after an economy drive.

These are before and after measurements made on the same 'individual'. More general cases can be devised particularly in testing procedures where pieces of raw material are divided in two and one part is subjected to a specified treatment and the other part is left untreated. The subsequent test results from treated and untreated parts form the two samples whose means are to be compared. Such sets of data are in the appropriate form for paired comparisons. It will be convenient to consider this test in terms of '*before*' and '*after*' data.

The test then is to decide whether or not there is a significant difference between the means of two sets of data; set b_i, the measurements taken 'before' and set a_i the measurements taken 'after'. The structure of the test is as follows:

(1) Adopt the null hypothesis that there is no difference between the 'before' and 'after' results.

(2) Adopt a significance level.

(3) Calculation: Because every member of the b sample has a corresponding member in the a sample one may work with the variable $(a - b)$ instead of the two variables a and b. Label this variable, for convenience, d, then:

$$d_1 = a_1 - b_1$$
$$d_2 = a_2 - b_2$$
$$\vdots \quad - \quad -$$
$$d_n = a_n - b_n$$
$$d = \frac{\Sigma d_i}{n} = \frac{\Sigma(a_i - b_i)}{n} = \frac{\Sigma a_i}{n} - \frac{\Sigma b_i}{n} = \bar{a} - \bar{b}$$

If the null hypothesis is true, then it is to be expected that $\bar{a} = \bar{b}$ or, equivalently, $d = 0$ but in practice d will differ slightly from zero.

Acceptance of the null hypothesis is then equivalent to accepting that d is insignificantly different from zero. d is the mean of n values $d_1, \ldots d_n$ and, if the null hypothesis is true, the true mean of the parent population of the ds must be 0. The variance of the parent population of the ds can be estimated as for any sample and from this the standard error of the means can be estimated as usual.

Hence the deviation of d from the mean 0 can be expressed in units of estimated standard error completely parallel to the calculation used in sampling theory. Because the calculation involves an *estimated population variance* the Normal distribution cannot be used and the '*t*' distribution is correctly used in its place. Hence:

$$t = \frac{d - 0}{\hat{\sigma}/\sqrt{n}} \text{ where } \hat{\sigma}^2 = \text{estimated population variance}$$

The number of degrees of freedom associated with '*t*' is $n - 1$. Critical values of '*t*' are tabulated (*see* Table T.2) where the columns are identified by possible significance levels and the rows by degrees of freedom.

(4) If the calculated value of '*t*' exceeds the tabulated value the result is signifi-

cant at that level and the null hypothesis is rejected. Otherwise the hypothesis is accepted.

Example. A pilot study is conducted on the effect of new forms of packaging for Product *X*. Monthly sales from 7 randomly selected shops are recorded firstly with the type *A* package and secondly with type *B*.

Shop No.	(Sales (dozens))	
	Packaging A	Packaging B
1	7	3
2	4	1
3	16	9
4	12	10
5	5	7
6	5	8
7	12	6

Does package *A* sell significantly different from package *B*?

Null hypothesis: There is no difference between the samples other than random sampling effects.

Significance level: 0·01 (or 1%)

Calculation: Results are of the paired comparison form hence adopt the variable *d*, the difference between the sales.

Shop No.	1	2	3	4	5	6	7
d	4	3	7	2	−2	−3	6

$$\Sigma d = 17 \quad \Sigma d^2 = 127 \quad d = \frac{17}{7} = 2\cdot43$$

$$\hat{\sigma}^2 = \left(\frac{127}{7} - \left(\frac{17}{7}\right)^2\right) \times \left(\frac{7}{6}\right) = 14\cdot28$$

$$\frac{\hat{\sigma}}{\sqrt{n}} = \sqrt{\frac{14\cdot28}{7}} = \sqrt{2\cdot04} = 1\cdot43$$

$$t = \frac{2\cdot43}{1\cdot43} = 1\cdot70 \text{ (6 degrees of freedom).}$$

From the table the critical value of $t = 3\cdot707$.

Conclusion: $1\cdot70 < 3\cdot707$, hence the hypothesis should be accepted.

On this evidence the difference in sales in favour of packaging *A* is insignificant.

6.6.2 DIFFERENCE BETWEEN TWO MEANS—GENERAL CASE

In this situation there is no relationship between individual members of the two samples and the samples are not necessarily of the same size.

Procedure: Data—Sample 1. n_1 values, mean m_1, sample variance s_1^2
Sample 2. n_2 values, mean m_2, sample variance s_2^2

(1) Adopt null hypothesis that both samples are random samples from a common population.
(2) Adopt a significance level.
(3) Calculation: Test first of all that sample variances are *not* significantly different using 'F' test. If these variances are significantly different a more complicated and approximate procedure may be adopted. This is described in Appendix E.

If s_1^2 is *not* significantly different from s_2^2:
Form a combined estimate of the population variance

$$\hat{\sigma}^2 = \frac{n_1 s_1^2 + n_2 s_2^2}{n_1 + n_2 - 2}$$

Calculate $t = \dfrac{m_1 - m_2}{\hat{\sigma}} \sqrt{\dfrac{n_1 n_2}{n_1 + n_2}}$

This has $n_1 + n_2 - 2$ degrees of freedom.
Compare this with the tabulated critical value as in the previous section.

Example. As in the previous example two new packaging forms are being test marketed, but this time simultaneously instead of sequentially. Eleven shops have been chosen and six of these sell type A packages, five sell type B packages. On the basis of the monthly sales figures can it be said that one is a better seller than the other?

Monthly sales (dozens)

Shop	Type A package	Shop	Type B package
1	14	7	8
2	7	8	10
3	12	9	5
4	17	10	6
5	10	11	8
6	12		

Null hypothesis: The difference in sales figures is due to chance, i.e. both sets of figures can be considered as samples from a single population or, equivalently, there is no real difference.
Significance level: 0·02 (or 2%)
Calculation: $n_A = 6$, $\Sigma x_A = 72$, $\Sigma x_A^2 = 922$
$\therefore m_A = 12\cdot00$, $s_A^2 = 9\cdot67$
$n_B = 5$, $\Sigma x_B = 37$, $\Sigma x_B^2 = 289$
$\therefore m_B = 7\cdot40$, $s_B^2 = 3\cdot04$

'F' test of variances: $\hat{\sigma}_A^2 = \dfrac{9 \cdot 67 \times 6}{5} = 11 \cdot 60$

$$\hat{\sigma}_B^2 = \dfrac{3 \cdot 04 \times 5}{4} = 3 \cdot 80$$

Variance ratio $= \dfrac{11 \cdot 60}{3 \cdot 80} = 3 \cdot 05$ with 5 and 4 degrees of freedom (as in Section 6.4).

Tabulated value at $1\% = 15 \cdot 5$, hence variances are *not* significantly different and may be combined to give a joint estimate of σ^2.

Combined estimate of $\sigma^2 = \dfrac{(6 \times 9 \cdot 67) + (5 \times 3 \cdot 04)}{6 + 5 - 2} = \dfrac{73 \cdot 22}{9} = 8 \cdot 14$

Hence:

$$t = \dfrac{12 \cdot 00 - 7 \cdot 40}{\sqrt{8 \cdot 14}} \sqrt{\dfrac{6 \times 5}{11}} \text{ (with 9 degrees of freedom)}$$

$$= \dfrac{4 \cdot 60}{2 \cdot 86} \times 1 \cdot 65 = 2 \cdot 66$$

The tabulated value is $2 \cdot 821$, hence the null hypothesis is retained, i.e. the evidence does not warrant the conclusion that the sample means are different if the probability of such a conclusion being wrongly rejected must be less than 2%.

Notes:

(1) If one had chosen to work at the 5% level the difference of means *would* have been significant because the tabulated value is $2 \cdot 262$. Thus on 5% of occasions values of t as high as $2 \cdot 262$ could occur by chance sampling variations but on only 2% of occasions could a value as high as $2 \cdot 821$ occur by chance alone.

(2) Both examples refer to the same type of question. Though the 'unpaired' comparison could be applied to the 'paired' comparison situation statistical experimental design considerations, which are outside the scope of this book, would show that the paired comparison test when possible is more 'powerful' than the non-paired test. Intuitively the reader will recognize that paired comparisons structure the results in such a way that the differences being considered can be attributed to the different conditions operating 'before' and 'after'. In the non-paired comparison the differences being considered include elements due to inherent differences amongst the subjects of the comparison. In this case sales from individual shops may be essentially high or low regardless of packaging. (An interesting discussion of statistical design is given in Duckworth [1].)

(3) Though statistical tests of this kind can establish the probable existence of a difference between means the decisionmaker must still decide whether or not the *magnitude and probability* of the difference warrant *action*.

(4) Care must be taken in relating the test procedure to the question being asked. The reader is referred to Section 6.8 where the distinction between one-tailed and two-tailed tests is discussed.

6.7 CHI-SQUARED TEST .

The tests so far considered have been applicable to continuous variables. The chi-squared test applies to discrete variables and is concerned with the question of whether or not the differences between an *observed* set of frequencies of occurrence of events and a *theoretically* expected set of frequencies are significant.

Procedure: Data—a set of observed frequencies $0_1, 0_2, \ldots 0_n$.

(1) Adopt null hypothesis that the differences between observed and theoretically expected frequencies are chance effects.

(2) Adopt a significance level.

(3) Calculate the theoretically expected frequencies $E_1, E_2, \ldots E_n$.

Take the difference between each pair of observed and expected frequencies $(0_1 - E_1), (0_2 - E_2) \ldots (0_n - E_n)$.

$$\text{Chi-squared } (\chi^2) = \sum_{i=1}^{n} \frac{(0_i - E_i)^2}{E_i}$$

(4) Compare the calculated value of χ^2 with the tabulated value using the appropriate number of degrees of freedom. If the calculated value is greater than the tabulated value reject the null hypothesis, otherwise accept it.

Example. Sloppo Breakfast Cereal is sold in three package sizes, Large Economy, Normal, and Small Handy. Experience has shown that these sell in the ratio $3:5:2$. Sales returns from a particular region show that 40, 45, and 15 cases respectively of the different package sizes have been sold. Is this pattern of sales significantly different from the previously established pattern?

Null hypothesis: There is no real difference between patterns
Significance level: 0·05 (or 5%)
Calculation:

Packet Sizes	L.E.	N	S.H.	Total
$O = Observed\ sales$	40	45	15	100
$E = Expected\ sales$	30	50	20	100
$\Delta = O - E$	10·0	5·0	5·0	
Δ^2/E	3·33	0·50	1·25	5·08

The total 5·08 is the value of chi-squared to be compared with a tabulated value. For this comparison the number of degrees of freedom is required and it is convenient in this test to view the question of degrees of freedom in a slightly different way. In calculating the expected frequencies (E_i) there is no calculation of a prior mean. It is however necessary that the totals of O_i and E_i should be equal. Hence when two of the three E_i values are determined the third is determined by the condition that $\Sigma E_i = 100$. The number of degrees of freedom is therefore two.

Consulting the chi-squared tables for 5% significance level and 2 degrees of freedom the critical value is 5·991. 5·08 is less than 5·991, therefore there is no evidence for rejection of the null hypothesis that only chance differences exist between the two patterns.

A particularly important use of the chi-squared test is in testing whether or not a particular set of observations conform to a specified probability distribution. This use allows discussion of two important points; a complication of the degrees of freedom determination and the step necessary when the *expected* frequency in a given class of observations is very small.

Example. Market survey data exists in the form of 400 interview record sheets. Each sheet contains 5 interview records and a count of the number of sheets containing 0, 1, 2, 3, 4, and 5 Brand X customers gives the Table shown below. Does this data conform satisfactorily to a Binomial distribution?

No. of Brand X customers/sheet	0	1	2	3	4	5
No. of sheets	64	146	129	50	9	2

Average number of Brand X customers/sheet

$$= \frac{(64 \times 0) + (146 \times 1) + (129 \times 2) + (50 \times 3) + (9 \times 4) + (2 \times 5)}{400}$$

$$= \frac{600}{400} = 1·50$$

Null hypothesis: The observed results come from a binomial distribution.
Significance level: 0·01 (or 1%)
Calculation: If the null hypothesis is valid the mean of the binomial distribution (np) is estimated to be 1·50. As $n = 5$, $p = \dfrac{1·50}{5} = 0·30$.

Using this value of p one can calculate the binomial probabilities of 0, 1, 2, 3, 4, and 5 Brand X customers/sheet.

$$P(r = 0) = {}^5C_0 \, (0·30)^0 \, (0·70)^5$$
$$P(r = 1) = {}^5C_1 \, (0·30)^1 \, (0·70)^4$$
$$P(r = 2) = {}^5C_2 \, (0·30)^2 \, (0·70)^3, \text{ etc.}$$

These evaluate to:

$r =$	0	1	2	3	4	5
P	0·168	0·360	0·309	0·132	0·028	0·002

As there are 400 sheets this in turn leads to the expected frequencies:

E 67·2   144·0   123·6   52·8   11·2   0·8

Bringing observed and expected frequencies together one obtains:

Brand X purchasers/sheet	0	1	2	3	4	5
O = Observed no. of sheets	64	146	129	50	9	2
E = Expected no. of sheets	67·2	144·0	123·6	52·8	11·2	0·8

At this point it can be recognized that the agreement is (qualitatively) very good and that a low value of chi-squared may be expected. One point must be dealt with first.

The chi-squared distribution is used here in a way reminiscent of the use of the normal distribution as an approximation to the binomial distribution. In that case the condition that $np > 5$ had to be observed. In this case a condition which must be observed is that none of the expected frequencies may be less than 5 otherwise the contribution of that component in the chi-squared total will be disproportionately large. (Different opinions are held as to the critical lower limit but the value of 5 quoted is frequently adopted.) In this problem the expected value of 0·8 is below the critical value and to satisfy the approximation condition class 5 must be coalesced with its neighbouring class 4 giving:

No. of Brand X users/sheet	0	1	2	3	4/5	*Total*
O = Observed	64	146	129	50	11	
E = Expected	67·2	144·0	123·6	52·8	12·0	

The chi-squared calculation is now carried out:

$\Delta = O - E$	3·2	2·0	5·4	2·8	1·0	
$\Delta^2/E =$	0·153	0·028	0·236	0·148	0·083	0·648

The question of degrees of freedom is a little more complex here. Originally there were 6 classes (0, 1, 2, 3, 4, 5). These were reduced to 5 by coalescing 4 and 5.

The mean of the binomial distribution was estimated from the data and hence the number of degrees of freedom is reduced by 1. The total of expected frequencies must equal the total of observed frequencies (as in the first example) and hence the number of degrees of freedom is reduced by a further 1.

Number of degrees of freedom = $6 - 1 - 1 - 1 = 3$

In general: Number of degrees of freedom = number of classes (after coalescing) − number of parameters calculated from data − 1.

The tabulated value of chi-squared for 3 degrees of freedom = 11·345.

Conclusion: 0·648 < 11·345, hence the null hypothesis is *not* discredited.

6.7.1 CONTINGENCY TABLES

It is often possible to cross-classify frequencies of occurrence according to two different criteria. The chi-squared test allows one to investigate the hypothesis that the two classification factors are independent.

Example. Sales of three lines of merchandise *A*, *B*, and *C* are recorded for five different regions. Is there evidence of particular preferences for certain lines in certain regions?

			Region			
Line	1	2	3	4	5	Totals
A	20	8	5	20	7	60
B	17	16	5	22	20	80
C	23	6	10	18	3	60
Totals	60	30	20	60	30	200

Null hypothesis: Different proportions of sales of *A*, *B*, and *C* in each region are due to chance fluctuations only, and not to particular regions having special preferences for particular lines.

Significance level: 1%

Calculation: If the null hypothesis is valid then the average and hence expected distribution of sales will be that shown in the right hand margin, i.e. 60:80:60 or 30% of *A*, 40% of *B*, and 30% of *C*. Total sales for each region are recorded in the bottom row. Proportioning these totals as 30%, 40%, 30% amongst the 3 lines gives the *expected table* below.

			Region			
Line	1	2	3	4	5	Total
A	18·0	9·0	6·0	18·0	9·0	60·0
B	24·0	12·0	8·0	24·0	12·0	80·0
C	18·0	9·0	6·0	18·0	9·0	60·0
Total	60·0	30·0	20·0	60·0	30·0	200·0

The differences between the two tables give the value of chi-squared just as before. Thus:

$$\text{Chi-squared} = \frac{(2\cdot0)^2}{18} + \frac{(1\cdot0)^2}{9} + \frac{(1\cdot0)^2}{6} + \frac{(2\cdot0)^2}{18} + \frac{(2\cdot0)^2}{9} + \frac{(7\cdot0)^2}{24} +$$

$$\frac{(4\cdot0)^2}{12} + \frac{(1\cdot0)^2}{6} + \frac{(2\cdot0)^2}{24} + \frac{(8\cdot0)^2}{12} + \frac{(5\cdot0)^2}{18} + \frac{(3\cdot0)^2}{9} +$$

$$\frac{(4\cdot0)^2}{6} + \frac{(0)^2}{18} + \frac{(6\cdot0)^2}{9} = 19\cdot262$$

In this case the number of degrees of freedom is determined by recognizing that both rows and columns in the expected tables must add up to the marginal totals. Thus only four entries in each row and two in each column could be made arbitrarily. There are therefore 4×2 degrees of freedom. In general for an $n \times m$ table of this kind there are $(n - 1) \times (m - 1)$ degrees of freedom.

The tabulated value of chi-squared for eight degrees of freedom and the 1% significance level is 20·09.

Conclusion: 19·26 < 20·09, therefore the null hypothesis is not rejected However the result is very nearly significant (and is significant at the 5% level— tabulated $\chi^2 = 15·507$) and therefore it would be desirable to collect more data (information) before drawing a final conclusion.

6.7.2 CHI-SQUARED WITH 1 DEGREE OF FREEDOM

One variation in the calculation procedure is necessary when the number of degrees of freedom is reduced to 1, a situation which occurs most commonly with 2×2 contingency tables. In such cases the following step is included in the calculation.

For each difference $\Delta_i = O_i - E_i$ reduce its magnitude by 0·5, *then* calculate

$$\sum \frac{\Delta_i^2}{E_i} = \chi^2.$$

The correction may be ignored when the difference between observed and expected values is less in magnitude than 0·5.

6.8 ONE- AND TWO-TAILED TESTS

The tests described in this chapter have been illustrated in their most naturally occurring forms. A variation which is applicable to most tests and which depends entirely on the form of the question being 'asked' of the data is that of setting up one- or two-tailed tests. The idea is most readily demonstrated in terms of the normal distribution.

In the motor-car tyre example (Section 6.5) the lives of tyres were normally distributed with mean 40,000 kilometres and standard deviation 3,000 km. One can imagine two distinct *types* of question which might be asked.

(a) What is the probability of a sample of 64 tyres having a mean life *greater than* 40,750 km? The calculation is:

Samples of size 64 will have their means normally distributed with mean = 40,000 and $S.E. = 375$.

Hence $\dfrac{40,750 - 40,000}{375} = 2$

From normal distribution tables 97·72% of the area lies to the left of the 40,750 km mark, hence the required probability is 2·28%. This is a one-tailed test because one is concerned with the proportion of area lying under one tail of the distribution (Figure 6.1).

(*b*) What is the probability of a sample of 64 tyres having a mean life 750 km or more *different from* the 'true' mean of 40,000? The calculation again takes one to the step:

$$\frac{40,750 - 40,000}{375} = 2$$

On consulting the normal distribution tables one is interested in the area of both the right- and left-hand tails 750 km (or 2 *S.E.*s) distant from the mean. The

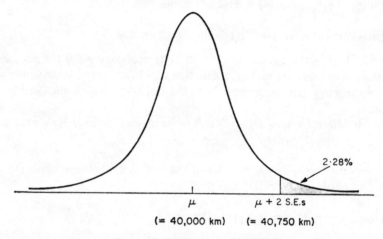

2·28%

μ μ + 2 S.E.s

(≈ 40,000 km)   (≈ 40,750 km)

Figure 6.1   Single-tail test situation

distribution is symmetrical and as 2·28% of area lies under the right-hand tail so 2·28% of area must be under the left-hand tail. Thus a total of 4·56% of the area is that much distant from the mean. Hence there is a probability of 4·56% of obtaining a sample of 64 tyres with mean life either greater than 40,750 or less than 39,250 (Figure 6.2).

This variant may require some careful thought by the reader. It should be pointed out that χ^2 and '*F*' tables are naturally constructed to meet one-tail tests while the '*t*' tables in this book are constructed to carry out two-tail tests.

Example. Consider the example in Section 6.6.2. The average Type *A* package sales are higher than the average Type *B* package sales (12·00 *v* 7·40). One might reasonably wish to ask whether or not sales of Type *A* packages are significantly higher than Type *B* package sales. (Compare this question carefully with that asked in Section 6.6.2.) In such a case a *one-tailed* test is appropriate and changes in the null hypothesis and in the use of the '*t*' table are required.

Null hypothesis: The apparently higher sales of Type *A* packages are due to chance alone, i.e. mean of *A* is not greater than the mean of *B*.

Significance level: 0·01 (or 1%) (as before).

Calculation: Calculated value of '*t*' is 2·66 with 9 degrees of freedom as before. The tabulated value required is that which cuts off 1% of the area under the '*t*' curve at one tail of the distribution. Because the '*t*' table is presented in a symmetrical form giving the critical value, for the α% significance level, which cuts

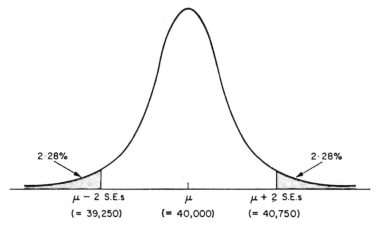

Figure 6.2   Two-tailed test situation

off $\frac{1}{2}$α% at the 'high' tail and $\frac{1}{2}$α% at the 'low' tail it is necessary in this case to consult the critical value for 9 degrees of freedom at the 2% level in order to obtain the desired value.

$$t_{2\%} = 2·821$$

2·66 < 2·821, hence the null hypothesis is not discredited, i.e. a difference of this magnitude could have been observed as a chance effect with a probability of greater than 1%.

Note. The only difference in using the table of critical values is that it is consulted for a significance level twice that stipulated.

6.9 SUMMARY

The significance testing procedures discussed here have as their common theme the attribution of any observed differences to chance sampling effects followed by a calculation of the probability of occurrence of a chance sampling effect of the observed magnitude. If this probability is unacceptably low then the difference observed is attributed to some 'systematic' mechanism rather than to chance.

Apparent differences in variance are tested using the '*F*' test. If the population variance is known apparent differences in means are tested using 'Area under the Normal Curve' tables. If the population variance is unknown and has to be estimated from sample data differences between means are tested using the '*t*' test.

Note that in all the above tests the assumption is implicit that the parent population is normally distributed. Procedures which do not invoke this assumption are described in Chapter 7. Differences between observed and expected (or predicted) frequencies of occurrences can be tested using the chi-squared test.

EXERCISES

1. Three interviewers carried out a door-to-door survey in a residential area. The following responses to a Yes–No question were obtained:

Interviewer	No. of responses	
	Yes	No
A	75	55
B	90	75
C	60	65

Is there evidence of interviewer bias in the sense that the nature of the response is in some way dependent on individual interviewers?

(*From University of Strathclyde*)

2. A company head office inspector wished to compare the strengths of cups of tea sold in two of its restaurants. He therefore took a sample from each of 8 cups of tea from one restaurant throughout one day and 8 cups of tea from the other restaurant the next day. Subsequent chemical analysis gave the following results:

Restaurant				Strength in g of solid/l.			
A	2·45	2·42	2·47	2·43	2·48	2·46	2·43	2·44
B	2·44	2·49	2·47	2·44	2·45	2·43	2·48	2·46

(*a*) Is there any real difference in strength between the two restaurants?

(*b*) If the company regulations state that the strength should be 2·46 g/l. is there any evidence that either restaurant is not complying with the regulation?

(*c*) Under what circumstances would the statistical tests you have used be invalid?

(*IOS, Part 2*)

3. Two works, *A* and *B*, provide the following data concerning absenteeism. Is this evidence that works *A* has a significantly higher absentee rate than works *B*?

	Works A	Works B
Average No. of absentees/1000 employees/wk	26	17
Variance of No. of absentees/1000 employers/wk	12	16
No. of weeks over which average was taken	8	10

(*From University of Strathclyde*)

4. In the packaging example in Section 6.6.1 show that had there been additional 2 shops included in the pilot study with sales as shown below, then the difference between packagings would have been significant at the 5% level.

Shop No.	Package A sales	Package B sales *(days)*
8	9	6
9	8	4

5. Examine Question 4 using the unpaired comparison test. What conclusion do you reach?

6. Re-calculate the market survey example in Section 6.7 using the data tabulated below. Show that this data can be represented satisfactorily by a Binomial distribution with $p = 0.35$.

No. of Brand X customers/sheet	0	1	2	3	4	5
No. of sheets	44	120	140	86	8	2

7. *Distribution-Free Significance Tests*

7.1 INTRODUCTION

The reader will have noticed in the discussion of Significance Tests (Chapter 6) that when '*F*' and '*t*' tests are used the parent population, from which the samples are assumed to be drawn, is stipulated as being *normally* distributed. Obviously this assumption cannot always hold and therefore alternative tests have been developed which do not require any assumption about the way in which the parent population of samples is distributed. Such tests are properly called *distribution-free* tests but are sometimes referred to as *non-parametric* tests. These terms are, strictly speaking, *not* equivalent but their usage as synonyms is quite common. There is continuing argument about the relative merits of classical significance tests and distribution-free significance tests and the authors of this book content themselves with two observations.

1. Where there are grounds for deeming the normal distribution assumption improbable distribution-free tests are safer.

2. Some readers may find the simpler arithmetic of the distribution-free tests attractive.

These observations apply to '*formal*' distribution-free tests. In the latter part of this chapter several '*informal*' distribution-free tests will be described. Such tests are sometimes referred to as 'quick and dirty' methods and have the advantage of lending themselves to 'back of envelope' calculation and subsequent interpretation without the need to refer to tables of critical values. It is emphasized that such tests should only be employed to give an indication of whether or not a more formal analysis might be rewarding and that decisions with appreciable cost consequences should *not* be based on such tests.

7.2 RANKING OF VALUES

The formal distribution-free tests presented here will involve manipulating the rank order number of values arranged in increasing order of magnitude. Thus:

Values	6·8	17·2	11·5	5·9	16·1	19·7	9·4	12·5	13·6	14·8
Ordered Values	5·9	6·8	9·4	11·5	12·5	13·6	14·8	16·1	17·2	19·7
Rank Order Number	1	2	3	4	5	6	7	8	9	10

It will be recognized immediately that transferring from the original values to the rank number simplifies the arithmetic because integers replace the original three-digit number. Also it will be recognized that in some sense information is being lost by making the transformation. One aspect of this may be appreciated by noting that though the rank order numbers are equally spaced on their 'scale' the original values are unequally spaced on the value 'scale', e.g. 6·8 is 0·9 units greater than 5·9 and is 2·6 units less than 9·4 but the rank number of 6·8, i.e. 2, is *midway* between the rank numbers of 5·9 and 9·4, i.e. 1 and 3.

7.3 TIED RANKS

In real data it is not uncommon for some values to be equal. Such values are said to be '*tied*'.

Example. 2·8, 3·6, 3·6, 3·6, 4·7, 5·9, 6·7, 6·7, 8·3, 9·9.

The values 3·6 are said to be triple tied and the values 6·7 double tied or, more commonly, just tied.

If these values are to be ranked, some tie-breaking procedure is necessary. Three simple possibilities exist:

(*a*) Obtain measurements to a further decimal place and hope thus to break the ties.

(*b*) Discard all tied values.

(*c*) Assign to each of the tied values the *average* rank order number which that set of ties would attract. (This is the method introduced in Chapter 10 in connection with rank correlation coefficients.)

Example.	2·8	3·6	3·6	3·6	4·7	5·9	6·7	6·7	8·3	9·9
	1	(2)	(3)	(4)	5	6	(7)	(8)	9	10
Rank Numbers	1	3	3	3	5	6	7·5	7·5	9	10

Of these (*a*) is not always practicable and (*b*) should only be adopted when the number of tied values is a very small proportion (say less than 5%) of the total number of values. Though (*c*) has the disadvantage of leading to non-integer rank numbers it is the simplest generally applicable tie-breaking procedure and is used throughout this chapter. It should be noted that tied ranks reduce the power of the tests and more sophisticated procedures for tie breaking have been developed. Readers interested in these alternative and theoretically more acceptable procedures are directed to Reference 2.

7.4 ARITHMETIC OF RANK ORDER NUMBERS

In Chapter 2 discussion of the properties of mode, median, and mean as measures of location or 'average' value led to the mean being identified as usually the most satisfactory measure. This view has been maintained, at least implicitly, in the subsequent chapters. 'Ranking' was first introduced in Chapter 2 as a step in the determination of the median of a set of values and the rank order based tests presented here should be considered as tests centred on sample *medians* in contrast to the corresponding classical significance tests which are centred on sample *means*. Some justification of the switch in emphasis from mean to median is obviously required. The simplest justification is that the dominant position of the mean in statistical theory can be attributed to the ease with which it may be manipulated mathematically. This advantage largely disappears when rank order numbers are substituted for proper values. These rank order numbers are basically a continuous sequence of integers and usually can be identified with 'the first N integers', i.e. the numbers $1, 2, 3 \ldots N - 1, N$.

The sum of the first N integers is given simply by:

$$\sum_{i=1}^{N} i = \tfrac{1}{2} N (N + 1)$$

Example. $N = 5$

$$\text{Sum} \sum_{i=1}^{5} i = 1 + 2 + 3 + 4 + 5$$
$$= 15$$

By formula: Sum $= \tfrac{1}{2} \times 5 \times 6 = 15$

7.5 BASIC TEST LOGIC

Distribution-free significance tests are set in the same logical framework as classical significance tests; that of setting up a null hypothesis, estimating the probability of actually observing values such as those obtained if that hypothesis is true and accepting or rejecting the hypothesis in the light of the probability value so obtained. In the distribution-free tests the mechanism for calculating probabilities is basically similar for each test and can conveniently be demonstrated in terms of the following test of whether or not two samples may be considered as coming from identically distributed populations.

7.6 WILCOXON'S SUMS OF RANKS TEST FOR SAMPLES OF EQUAL SIZE

Example. A new product is being test marketed in two types of package, Type *A* and Type *B*. Initial interest is in retailers' reaction and from twenty randomly selected shops of similar type ten are offered Type *A* packages and ten Type *B* packages. The size (in dozens) of orders placed are as shown below:

| Type A | 8 | 9 | 10 | 11 | 13 | 14 | 17 | 18 | 20 | 26 |
| Type B | 9 | 12 | 15 | 16 | 18 | 18 | 19 | 22 | 24 | 27 |

Test procedure:

Null hypothesis: There is no effect attributable to package type. Both sets of values are random samples from identically distributed populations whose distribution form is not assumed.

Significance level: 0·05 (say)

Probability calculation: Combine both sets of order values into one group and rank the values within that group. Reassign the rank values obtained to the original package types. This is shown in Table 7.1.

TABLE 7.1

Ranking of values considered as a single group

Value	8	9	9	10	11	12	13	14	15	16	17	18	18	18	19	20	22	24	26	27
Package type	A	B	A	A	A	B	A	A	B	B	A	A	B	B	B	A	B	B	A	B
Rank number	1	2·5	2·5	4	5	6	7	8	9	10	11	13	13	13	15	16	17	18	19	20

(The reader is asked to note that tie-breaking has been employed for the doubly tied 9s and the trebly-tied 18s and that this has not affected the range of integers used as rank numbers, i.e. $N = 20$.)

Reconstructing the package-type groups gives:

Sum of
rank numbers
for Type A $= 1 + 2·5 + 4 + 5 + 7 + 8 + 11 + 13 + 16 + 19 = 86·5$
Sum of
rank numbers
for Type B $= 2·5 + 6 + 9 + 10 + 13 + 13 + 15 + 17 + 18 + 20 = 123·5$

(A simple check at this point shows that:

Sum of integers from 1–20 $= \frac{1}{2}(20) \times (21) = 210 = 86·5 + 123·5$.)

The next step in the calculation is based on the argument that if the low and high *values* are uniformly spread between the two types of sample as postulated, by the null hypothesis, then the sum of *ranks* in both samples should be approximately equal and should have values close to $210/2 = 105$, whereas if the high values tend to be found preferentially in one sample the sum of ranks of that sample will be higher than the average of value of 105. In the example being considered the problem is then condensed to determining whether or not the value of 86·5 can be considered as being that much less than the theoretical mean of 105 because of random sampling effects alone. (The value of 123·5 could equally well be considered in this case of equal sample sizes.)

Tables of critical values for this test are published as they are for every statistical test and the reader may find these in advanced texts, but where $N \geqslant 20$ the ubiquitous Table of Areas under the Normal Curve may be used to give a good approximation to the required probability. Note that the use of this argument does *not* compromise the original refusal to make assumptions about the distribution of the parent population.

To use the Normal Curve Tables it is necessary to know the mean and standard error of the distribution being considered. The appropriate formulae are:

$$\mu = \tfrac{1}{4}(N)(N+1) = 105$$

$$\hat{\sigma} = \sqrt{\frac{N^2(N+1)}{48}} = 13 \cdot 2$$

The Table is then used in the well-known way:

$$z = \frac{105 \cdot 0 - 86 \cdot 5 - 0 \cdot 5}{13 \cdot 2} = \frac{18 \cdot 0}{13 \cdot 2} = 1 \cdot \dot{3}6$$

The probability of getting a value more than $1 \cdot 36$ standard deviations from the mean by chance alone is $0 \cdot 1738$. This is greater than the adopted significance level of $0 \cdot 05$, hence the null hypothesis is not rejected. Note that a two-tailed test is used here.

7.7 WILCOXON'S TEST FOR UNEQUAL SAMPLE SIZES

Designate the sample sizes to be n and m so that $N = n + m$. In this case the only change in the calculation is that the expected sum of ranks in a sample is given by:

$$\tfrac{1}{2}n(N+1) \text{ or } \tfrac{1}{2}m(N+1)$$

and the standard error of the sample sum of ranks is given by:

$$\hat{\sigma} = \sqrt{\frac{nm(N+1)}{12}}$$

Example. Suppose that twenty-two retailers had been selected for the test marketing exercise and that twelve had been offered Type *A* packages and ten Type *B* packages. The sizes, in dozens, of orders placed is as shown:

Type A	7	8	9	9	10	11	13	14	17	18	20	26
Type B	9	12	15	16	18	18	19	22	24	27		

Test procedure:

Null hypothesis: Both sets of values are random samples from identically distributed populations of order sizes.

Significance level: $0 \cdot 05$

Probability calculation: Combine the values into a single ranked group.

Value	7	8	9	9	9	10	11	12	13	14	15	16	17	18	18	18	19	20	22	24	26	27
Package Type	*A*	*A*	*B*	*A*	*A*	*A*	*A*	*B*	*A*	*A*	*B*	*B*	*A*	*A*	*B*	*B*	*B*	*A*	*B*	*B*	*A*	*B*
Rank Number	1	2	4	4	4	6	7	8	9	10	11	12	13	15	15	15	17	18	19	20	21	22

Sum of rank numbers
for Type A
$$= 1 + 2 + 4 + 4 + 6 + 7 + 9 + 10 + 13 + 15 + 18 + 21 = 110$$

Sum of rank numbers
for Type B
$$= 4 + 8 + 11 + 12 + 15 + 15 + 17 + 19 + 20 + 22 = 143$$

(Check: Sum of integers from 1 to 22 $= \frac{1}{2}(22)(23) = 253 = 110 + 143$.)

Either sum of ranks may be tested against the expected sum for that group using the normal distribution (Table T.1). Applying the calculation to the Type A sum of ranks gives:

Observed Sum of ranks $= 110$. $n = 12$.

Expected Sum of ranks (under
the null hypothesis) $= \frac{1}{2} \times 12(23) = 138$

Standard error of Sum of ranks $= \sqrt{\dfrac{12 \times 10 \times (23)}{12}} = \sqrt{230} = 15 \cdot 2$

$$\therefore z = \frac{138 - 110 - 0 \cdot 5}{15 \cdot 2} = \frac{27 \cdot 5}{15 \cdot 2} = 1 \cdot 8$$

From Table T.1 the probability of getting a rank sum more than $1 \cdot 8$ standard deviations from the expected value under the null hypothesis is $0 \cdot 0718$. This is greater than $0 \cdot 05$, hence the null hypothesis is not rejected.

7.8 WILCOXON'S SIGNED RANK TEST

This is an alternative to Student's 't' test as applied to paired comparison data, i.e. to 'before' and 'after' data.

Example. Volumes of sales are recorded in twelve selected marketing areas before and after a change in the company's national advertising policy. Is there a significant increase in sales?

Market area	1	2	3	4	5	6	7	8	9	10	11	12
Before	19	28	11	34	32	14	40	23	28	21	18	20
After	16	35	13	36	32	19	50	29	36	20	22	29
Difference $(A - B)$	−3	+7	+2	+2	0	+5	+10	+6	+8	−1	+4	+9

Test procedure:

Null hypothesis: There is no significant difference between the 'before' and 'after' results. Both sets are random samples from identically distributed populations.

Significance level: 0·05
Probability calculation:
Order the differences according to their *magnitude* ignoring for the moment the + and − signs

Ordered Differences	0	1	2	2	3	4	5	6	7	8	9	10
Sign	−	−	+	+	−	+	+	+	+	+	+	+
Rank number	1	2	3·5	3·5	5	6	7	8	9	10	11	12

Note: 0 is given a − sign here because it 'supports' the null hypothesis. In each case the sign attached to zero must be determined with reference to the null hypothesis. Some authors (e.g. reference 2) suggest ignoring zero differences.

Sum of + ranks = 70: Sum of − ranks = 8.
(Check: 1/2 (12) (13) = 78).

The small value of the sum of ranks of the negative differences is due to the combined effect of their being small in number and in magnitude. Both factors tend to discredit the null hypothesis. This again can be tested using the normal distribution tables and the following formulae are required:

Expected sum of ranks $= n(n + 1)/4$
Standard error of sum of ranks $= \sqrt{[n(n + 1)(2n + 1)/24]}$.
Expected sum of ranks $= 12 \times 13/4 = 39$
SE of sum of ranks $= \sqrt{[12 \times 13 \times 25/24]} = 12\cdot747$
Hence $z = \dfrac{39 - 8 - 0\cdot5}{12\cdot747} = \dfrac{30\cdot5}{12\cdot747} = 2\cdot39$

From the table the corresponding probability is 0·0084. This is less than 0·05 hence the null hypothesis is rejected. Note that this is a one-tailed test since interest is centred on a significant *increase* in sales and not on *difference* in sales.

7.9 KRUSKAL AND WALLIS TEST

This test is an alternative to the one-way analysis of variance test (*see* Chapter 12) and tests for significant differences amongst three or more samples of data, the samples not necessarily being of equal size. Essentially the test procedure is an extension of the sum of ranks test already described.

Example. The orders received from a randomly selected number of comparable shops in Birmingham, Glasgow, and Manchester are as shown. Is there a significant difference in product acceptance amongst these cities?

No. of items ordered

Birmingham	24	31	42	43	45
Glasgow	21	23	30	38	39
Manchester	31	40	41	44	

Test procedure:

Null hypothesis: There is no significant difference. The values recorded are random samples from identically distributed populations.

Significance level: 0·05

Probability calculation:

(*a*) Group all the data and rank regardless of city of origin.

Values	21	23	24	30	31	31	38	39	40	41	42	43	44	45
City	G	G	B	G	M	B	G	G	M	M	B	B	M	B
Rank numbers	1	2	3	4	5·5	5·5	7	8	9	10	11	12	13	14

(*b*) Reclassify the rank numbers according to city and calculate the rank sum for each.

						Sum of Ranks
Birmingham	3	5·5	11	12	14	45·5
Glasgow	1	2	4	7	8	22·0
Manchester	5·5	9	10	13		37·5

(Check: Sum of integers 1 to 14 = 105 = 45·5 + 22·0 + 37·5.)

(*c*) For each city calculate the square of the sum of ranks and divide by the number of rank numbers in that class to get the mean square.

	Birmingham	*Glasgow*	*Manchester*
Number of ranks	5	5	4
Sum of ranks	45·5	22·0	37·5
Square of sum of ranks	2,070·25	484·0	1,406·25
Mean square	414·05	96·80	351·56

(*d*) Add these mean squares (862·41 in this case), then calculate

$$H = \frac{12}{N(N+1)} \text{(sum of mean squares)} - 3(N+1)$$

where N = Total number of values

This gives in this example:

$$H = \left(\frac{12}{14 \times 15} \times 862·41 \right) - 3 \times 15$$

$$= 49·28 - 45 = 4·28$$

(*e*) Consult χ^2 tables using a number of degrees of freedom one less than the number of *classes* of value being examined. In this case degrees of freedom = 3 − 1 = 2. For a significance level of 0·05 the critical upper value is 5·991 (from Table T.3). 4·28 is less than this, hence the null hypothesis is accepted.

The rationale for this test hinges on the fact that if the null hypothesis were true and if the classes contained equal numbers of values the expected rank totals for each class would be the same and the sum of squares of these totals would be less than for any *unequal* partitioning of the overall rank sum over the classes.

Example. Suppose fifteen values are being considered, five to each of three classes.

$$N = 15. \text{ Overall sum} = \tfrac{1}{2} \times 15 \times 16 = 120.$$

Equal partitioning gives each class a rank sum of 40. Sum of squares over all 3 classes $= 40^2 + 40^2 + 40^2 = 4,800$. If the partitioning is unequal, say 35, 35, 50, then sum of squares over all 3 classes $= 35^2 + 35^2 + 50^2 = 4,950$. For any other unequal partitioning a value greater than 4,800 will be found.

Division by the number of values in each class allows the chi-squares distribution to be used as a source of critical values and permits unequal numbers in each class to be considered.

7.10 'QUICK AND DIRTY' METHODS

Like the *formal* distribution-free tests the *'informal'* tests achieve their arithmetic simplicity by sacrificing some of the information contained in the values being tested. In the most general type of test to be described the actual values are discarded and the test is based on the number of values in one set which are greater or less than corresponding values in another set. Once again it must be emphasized that the results of these tests should be used as guidelines as to whether or not more formal analyses should be carried out and should not be used as a basis for critical decision-making.

7.11 2 . $\sqrt{N}$ TEST

This is a ubiquitous test demonstrated most readily in terms of a paired comparison situation which might finally be treated by either Student's 't' test or Wilcoxon's Signed Ranks test. Consider the following set of 'before' and 'after' data from which one wishes to establish whether or not the 'after' values are significantly greater than the 'before'.

Before	16	14	17	18	12	11	17	14	16	18	14	15	18	14	16	13
After	20	15	16	19	15	16	16	15	17	20	15	15	20	14	17	12
Sign of difference	+	+	−	+	+	+	−	+	+	+	+	0	+	0	+	−

Test procedure:

Null hypothesis: Both sets of values are random samples from the same population.

Significance level: 0·05. (In this form the test requires this significance level to be adopted. A note as to how different significance levels may be used is included at the end of this example.)

Probability calculation:

(*a*) Examine the relative magnitudes of values in each pair.

If $A > B$ record a $+$ sign
If $A < B$ record a $-$ sign
If $A = B$ record 0.

(*b*) Ignore all pairs tagged 0. This reduces the number of pairs being considered in the above example to 14.

(*c*) Count the number of $+$'s and $-$'s (11 and 3 in this example).

(*d*) Divide the difference of these values by twice the square root of their sum. If this ratio is greater than 1·0 the result is significant at the 0·05 level and the null hypothesis should be discarded. In this case:

$$\text{Ratio} = \frac{11 - 3}{2 \cdot \sqrt{(11 + 3)}} = \frac{8}{2 \cdot \sqrt{14}} = \frac{8}{7 \cdot 48} > 1 \cdot 0$$

Hence the 'after' values are significantly greater than the 'before' values at the 0·05 level.

Note. If one desires to work at other significance levels the necessary changes are:

For 0·01 (1%) level: Divide difference by 2·6 . $\sqrt{Sum}$
For 0·001 (0·1%) level: Divide difference by 3·3 . $\sqrt{Sum}$

If the ratios so obtained are greater than 1·0 the result is significant at that level.

7.12 TREND TEST: $2 . \sqrt{N}$ APPLICATION

Divide the sequence of values being tested into three approximately equal-sized groups of which the 1st and 3rd group *must* be of equal size. Compare the first value in the 1st group with the first value in the 3rd group assigning a $+$ or $-$ as one is greater or less than the other and a 0 if they are equal. Continue for all corresponding pairs in these groups. Treat the resulting $+$'s and $-$'s exactly as in the previous test.

Example. Does the following sequence of values show a trend?

17, 18, 21, 20, 17, 20, 23, 19, 18, 19, 20, 20, 21, 19, 25, 23, 21, 18, 19, 20, 22, 18, 26, 26, 24, 25.

There are 26 values in the sequence. The nearest approach to equal groupings which can be achieved is $9 + 8 + 9$.

Group 1.   17, 18, 21, 20, 17, 20, 23, 19, 18
Group 2.   (19, 20, 20, 21, 19, 25, 23, 21)
Group 3.   18, 19, 20, 22, 18, 26, 26, 24, 25
         +   +   −   +   +   +   +   +   +
       No. of + signs = 8
       No. of − signs = 1

$$\text{Ratio} = \frac{8 - 1}{2 \cdot \sqrt{(8 + 1)}} = \frac{7}{2 \cdot \sqrt{9}} = \frac{7}{6} > 1 \cdot 0$$

Hence there is a trend significant at the 0·05 significance level.

7.13 CORRELATION TEST: $2 \cdot \sqrt{N}$ APPLICATION

Note: Correlation is discussed in Chapter 10.

The data consists of a set of paired values (x_i, y_i).
Calculate the average values of x and y and compare each actual number pair with this hypothetical pair $(\bar{x}, \bar{y})$, where $\bar{x}$ = mean of x_is and $\bar{y}$ = mean of y_is.
If for any pair $(x_i y_i)$

     $x_i > \bar{x}$ *and* $y_i > \bar{y}$

or $x_i < \bar{x}$ *and* $y_i < \bar{y}$, assign to that pair a '+'.
If, however, $x_i > \bar{x}$ and $y_i < \bar{y}$
or $x_i < \bar{x}$ and $y_i > \bar{y}$, assign to that pair a '−'.
If either $x_i = \bar{x}$ or $y_i = \bar{y}$, assign to that pair a '0'.
Treat the +'s and −'s exactly as in the previous examples.

Example. Do the following pairs of values show a significant correlation?

X	5	6	6	8	9	10	12	12	14	14	15	21
Y	2	5	3	5	7	7	6	8	10	11	16	16
	+	+	+	+	+	+	−	0	+	+	+	+

$\bar{x} = 11$, $\bar{y} = 8$ hence the +'s and −'s assigned.
There are 10 + and 1 − signs here.

$$\text{Ratio} = \frac{10 - 1}{2 \cdot \sqrt{11}} = \frac{9}{6 \cdot 6} > 1 \cdot 0.$$

The correlation is therefore significant at the 5% level. This result is also significant at the 1% level. Using the multiplier of 2·6 given in the note earlier one gets:

$$\text{Ratio} = \frac{10 - 1}{2 \cdot 6 \cdot \sqrt{11}} = \frac{9}{8 \cdot 58} > 1 \cdot 0.$$

The assignment of +'s, −'s, and 0's in this test is rather clumsy to express in words. Reference to Figure 7.1 may help the reader. The points defined by the pairs of numbers are plotted on a graph and vertical and horizontal subsidiary axes are drawn through the mean values.

+ signs are assigned to points in one pair of diagonally opposite quadrants and − signs to points in the other pair of quadrants. 0 signs are assigned to any point falling on either subsidiary axis.

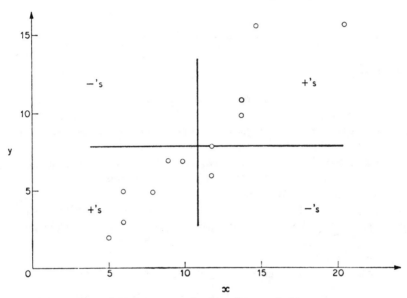

Figure 7.1   Diagram for 2 . $\sqrt{N}$ correlation test

7.14 TUKEY'S TEST

In the 2 . $\sqrt{N}$ tests given the need to compare values has necessitated equal numbers of values in each of the samples being compared. Tukey's test is applicable to two groups of different size, though to use the significance level values quoted here the difference in sample sizes should not be greater than ten. The only other condition required is that one sample should contain the highest value of the combined set and the *other* sample the lowest value.

The test simply involves counting the number of values in one sample greater than all other values in the other sample and adding to this the number of values in the other sample less than all values in the first sample. The sum thus obtained is significant according as:

Sum > 7. Significant at 0·05 level
Sum > 10. Significant at 0·01 level
Sum > 13. Significant at 0·001 level

Example. Do the following sales returns from regions *B* and *G* suggest that sales in *G* are significantly higher than in *B*?

    B   26, 28, 29, 42, 51, 54, 59, 62, 62, 68, 70, 73
    G   30, 35, 36, 47, 53, 68, 74, 75, 76, 80, 80, 84, 86, 86

    No. of values in *G* greater than largest in *B* (73) = 8
    No. of values in *B* less than smallest in *G* (30)   = 3
      Sum = 11.

The difference is therefore significant at the 1% level.

7.15 SUMMARY

Classical significance tests require that assumptions about samples being drawn from a common population include the condition that that population is normally distributed. This restriction can be avoided by using distribution-free tests but only at the cost of sacrificing some of the information contained in the values being tested. The relative advantage of this trade-off is a matter for debate though the attraction of the less laborious arithmetic required by certain distribution-free tests is beyond question.

In this chapter a distinction is made between formal distribution-free tests which are, in many eyes, respectable alternatives to certain classical significance tests and informal or 'quick and dirty' methods which may be used in rough and ready assessment of data.

The formal tests presented relate to classical tests as follows:

Wilcoxon's Sum of Ranks Test : Student's '*t*' test of sample means.
Wilcoxon's Signed Ranks Test : Student's '*t*' test applied to paired comparison data.
Kruskal and Wallis Test : Analysis of Variance on one-way classified data.

In all these cases the use of Tables of Area under the Normal Curve is suggested as the source of critical values. This presupposes that samples are not too small and with values of $N \geq 20$ the approximation is usually acceptable. Some critical tables for these tests for use with small samples will be found in Reference [3].

EXERCISES

1. Could either of the following pairs of samples be considered as coming from identically distributed populations?

(*a*) *A* values: 9·6, 6·8, 7·2, 13·1, 12·8, 16·4, 9·3, 14·2, 18·7, 16·1, 12·6.
    B values: 5.2. 9·7, 7·0, 10·2, 13·2, 6·6, 15·7, 9·9, 16·3, 14·1, 8·5.

(b) *A* values: 9·6, 6·8, 7·2, 5·2, 7·0, 9·7, 9·3, 14·2, 14·1, 8·5, 9·9
    B values: 13·1, 12·8, 16·4, 10·2, 13·2, 6·6, 15·7, 16·3, 18·7, 12·6, 16·1.

2. Could the following pairs of samples be considered as coming from identically distributed populations?

(a) *X* values: 5·3, 0·6, 1·8, 2·5, 8·8, 8·4, 6·1, 3·8, 7·0, 4·1, 4·7, 3·8.
    Y values: 1·7, 1·8, 9·0, 9·7, 9·4, 6·9, 9·6, 3·0, 7·1, 8·5

(b) *U* values: 0·6, 1·7, 1·8, 2·5, 3·0, 3·8, 3·8, 4·7, 5·3, 6·1, 6·9, 8·8.
    V values: 1·8, 4·1, 7·0, 7·1, 8·4, 8·5, 9·0, 9·4, 9·6, 9·7

3. After a change of advertising agency the following changes in regional sales are observed. What conclusion do you draw?

Changes in sales volume (£'000s)
+30, −10, +20, +15, −6, +40, +15, +17, −12, +8, +10, +35, −14,
+4, +3, +2

4. Sales returns from randomly selected shops in four cities give the following values (£'000s) for a specified product. Does this indicate a significant difference in product acceptance amongst cities?

Birmingham	25	40	35	30	36	
Glasgow	28	34	31	37		
Manchester	42	44	33	22	29	
Liverpool	19	24	39	45	23	41

5. Apply the $2 \cdot \sqrt{N}$ test to the data in Question 3. What conclusion do you draw?

6. Apply the Tukey test to the data in Questions 1 and 2. What conclusions do you draw?

7. Using the $2 \cdot \sqrt{N}$ test examine the correlation between the following sets of numbers:

(a) *x:* 1·5, 6·1, 3·1, 3·2, 7·0, 4·7, 4·8, 6·9, 6·3, 2·6, 6·9, 4·5
    y: 3·4, 12·5, 6·5, 6·0, 13·0, 10·0, 10·0, 14·5, 12·9, 4·0, 12·0, 8·0

(b) *x:* 1·0, 3·5, 6·0, 6·5, 1·5, 2·0, 7·0, 4·0, 5·5, 2·5, 3·0, 5·0, 4·5
    y: 4·0, 14·0, 25·0, 30·0, 5·0, 8·0, 29·0, 12·0, 21·0, 4·0, 15·0, 21·0, 7·0.

8. *Conditional Probability and Bayes' Theorem*

8.1 INTRODUCTION

In Section 3.4.1 simple multiplication of probabilities was stated to be permissible given that the events concerned were independent. Examples of non-independent events were suggested in that chapter but to emphasize the idea of non-independence the following example is offered.

Example. Suppose that a market survey estimated that 16% of the adult population are cigar smokers. A breakdown of this figure by sex might give the results of Table 8.1.

TABLE 8.1 **Cigar smokers?**

	Yes	No	
Adult male	290	660	950
Adult female	30	1,020	1,050
	320	1,680	2,000

The 16% figure is obtained by dividing the number of cigar smokers (320) by the total number interviewed (2,000). Obviously the incidence of cigar smoking in adult males is considerably higher (290/950 = 30·53%) and the incidence of cigar smoking in adult females much lower (30/1050 = 2·86%) than the 16% figure. Both figures are relative frequency estimates of the probability of an individual being a cigar smoker *given that the individual's sex is specified.*

In formal notation one can write:

$P(X$ is a cigar smoker | Adult Male$) = 0.3053$
$P(X$ is a cigar smoker | Adult Female$) = 0.0286$

where the vertical stroke (|) is read as 'given that X is'.

Such probabilities are termed *conditional probabilities* because the value specified is only valid if the condition stipulated by the vertical stroke is satisfied.

In general one can write $P(A|B) = x$ where this is to be read as 'The probability of A occurring *given that B is true* equals x'.

146

8.2 MANIPULATION OF CONDITIONAL PROBABILITIES

Using the notation C for cigar smoker, M for male, etc., the following probability estimates can be written down from the Table 8.1:

$$P(C) \quad = 320/2000 = 0 \cdot 160$$
$$P(M) \quad = 950/2000 = 0 \cdot 475$$
$$P(C|M) = 290/950 \quad = 0 \cdot 305$$
$$P(M|C) = 290/320 \quad = 0 \cdot 906$$

From these it can be seen that:

$$P(M|C) \times P(C) = \frac{290}{320} \times \frac{320}{2,000} = \frac{290}{2,000}$$

and

$$P(C|M) \times P(M) = \frac{290}{950} \times \frac{950}{2,000} = \frac{290}{2,000}$$

Inspection of the table shows that, using the relative frequency interpretation, $290/2,000 = P(MC)$ is the probability of an individual being both *male* and a cigar smoker.

[The reader is asked to consider carefully the distinction between $P(CM)$, $P(M|C)$, and $P(C|M)$. It may be of assistance to imagine that one is concerned with making statements about the sex and/or cigar-smoking characteristics of the next adult entering a room (say). If no information about either characteristic is provided the probability that the individual is male *and* a cigar smoker is given by $P(CM)$ (or equivalently by $P(MC)$ the order of characteristics in *this case* being immaterial.) If one is told that the individual *is a cigar smoker* then the probability that he is male is given by $P(M|C)$, and if one is told the individual *is male* then the probability that he is a cigar smoker is given by $P(C|M)$.]

It can be seen that a cigar smoker *must* be either male or female, 'male' and 'female' being in probability theory terms *mutually exclusive* and *exhaustive* states. Hence one can write:

$$P(C) = P(CM) + P(CF)$$
But $P(CM) = P(C|M) \times P(M)$ and $P(CF) = P(C|F) \times P(F)$, hence
$$P(C) = P(C|M) \times P(M) + P(C|F) \times P(F)$$

This last result can be confirmed by substituting the numerical values from Table 8.1 giving:

$$\frac{320}{2,000} = \frac{290}{950} \times \frac{950}{2,000} + \frac{30}{1,050} \times \frac{1,050}{2,000}$$
$$= \frac{290}{2,000} + \frac{30}{2,000}$$

In general terms one can write the principal results of conditional probability theory as:

$$P(A|B) \times P(B) = P(AB) = P(B|A) \times P(A)$$

and

$$P(A) = P(AX_1) + P(AX_2) + P(AX_3) + \ldots P(AX_n)$$

where $X_1, X_2, \ldots X_n$ are *mutually exclusive* and *exhaustive states* with which the occurrence of A must be associated.

This in turn expands to:

$$P(A) = P(A|X_1) \times P(X_1) + P(A|X_2) \times P(X_2) + \ldots + P(A|X_n) \times P(X_n)$$

The reader is asked to note that in general $P(A)$ is *not* equal to $P(A|B)$. If the equality *does* hold, then A and B are independent.

Example. The probability of drawing an ace from a pack of playing cards is $4/52 = 1/13$. The probability that a card drawn is an ace *given that it is a red card* is $2/26 - 1/13$, i.e. $P(\text{Ace}) = 1/13$ and $P(\text{Ace}|\text{Red card}) = 1/13$. This implies that the colour and face value of playing cards are independent and, as all suits are identically constituted, this is of course true.

It is often useful to consider a conditional probability value as a probability value modified in the light of information provided, and, as will be shown in subsequent sections, a systematic procedure for updating probabilities in the light of information is available and useful.

8.3 BAYES' THEOREM

Situations frequently occur in which it is possible to calculate or estimate $P(A|B)$ say but interest is really in obtaining a value for $P(B|A)$. This can often be achieved using the relationship given in the previous section, namely:

$$P(A|B) \times P(B) = P(AB) = P(B|A) \times P(A)$$

The middle term may be omitted to obtain the direct equality:

$$P(A|B) \times P(B) = P(B|A) \times P(A)$$

This can be rearranged to give:

$$P(B|A) = \frac{P(A|B) \times P(B)}{P(A)}$$

In its most usual form the occurrence of A is associated with a number of mutually exclusive and exhaustive events $X_1, X_2 \ldots X_n$. The probabilities of A occurring given that any of the $X_1 \ldots X_n$ has occurred is known. Interest is focused on obtaining the probability of (say) X_i occurring given that A is known to have occurred. This can be written, following the last equation:

$$P(X_i|A) = \frac{P(A|X_i) \times P(X_i)}{P(A)}$$

Because A is necessarily associated with the events $X_1 \ldots X_n$ this can be rewritten, using the result of the previous section, as:

$$P(X_i|A) = \frac{P(A|X_i) \times P(X_i)}{P(A|X_1) \times P(X_1) + P(A|X_2) \times P(X_2) + \ldots P(A|X_n) \times P(X_n)}$$

This expression is known as *Bayes' Theorem* and provides the logical basis for a useful, if sometimes contentious, approach to a variety of statistical problems. It should be noted that the equation may be looked on as a means of modifying the value $P(X_i)$ on the right hand side of the equation to the value $P(X_i|A)$, in the light of the information that A has occurred. $P(X_i)$ may be referred to as the *prior probability* of occurrence of X_i and $P(X_i|A)$ as the *posterior probability* of occurrence of X_i. 'Prior' and 'posterior' imply of course knowledge available 'before' and 'after' the occurrence of A.

8.4 'UPDATING' PROBABILITIES

In Section 6.2 the problem of deciding whether or not a coin should be deemed 'normal' or 'double-headed' on the basis of the reported results of tosses is used to illustrate the logical framework of significance testing. Bayes' theorem allows a different approach to this problem which demonstrates clearly the idea of updating probabilities in the light of information acquired.

Though the problem is artificial it serves very well to illustrate the mechanism and the arguments of the Bayesian approach. A coin may either be 'normal' (n) or 'double-headed' (dh). The decision-maker is not allowed to inspect the coin but can have made available to him the results of successive tosses of the coin. If the decision-maker had not access to experimental evidence of this kind he would hold nevertheless some opinion of the relative probabilities of the coin being 'n' or 'dh'. This opinion might be largely subjective based on his accumulated experience of coins of each type or could be based on a formal examination of a large number of coins. Suppose in this case that his opinion is formed subjectively and that he assigns probabilities as follows:

P (coin is normal) $= P(n) = 0.99$
P (coin is double-headed) $= P(dh) = 0.01$

The decision-maker then calls for the coin to be tossed 'k' times. If the result of these tosses is k heads how is he to modify his opinion as to the relative probabilities of the coin being 'normal' or 'double-headed'? In conditional probability terms he is seeking values for:

$P(n|k$ heads$)$ and $P(dh|k$ heads$)$

$P(n) = 0.99$ and $P(dh) = 0.01$ are reasonable prior likelihoods. If then a sequence of k heads is reported for independent tosses of the same coin the modified probability required is:

$$P(n|k \text{ 'heads'}) = \frac{P(k \text{ 'heads'}|n) \times P(n)}{P(k \text{ 'heads'}|n) \times P(n) + P(k \text{ 'heads'}|dh) \times P(dh)}$$

$P(k \text{ 'heads'}|n) = (\frac{1}{2})^k.$ $P(k \text{ 'heads'}|dh) = 1.00.$

Suppose $k = 2$.

$$P(n|2 \text{ heads}) = \frac{0.25 \times 0.99}{(0.25 \times 0.99) + (1.00 \times 0.01)} = \frac{0.2475}{0.2475 + 0.0100}$$

$$= \frac{0.2475}{0.2575} = 0.961$$

Suppose $k = 3$.

$$P(n|3 \text{ heads}) = \frac{0.125 \times 0.99}{(0.125 \times 0.99) + (1.00 \times 0.01)} = \frac{0.12375}{0.12375 + 0.01000}$$

$$= \frac{0.12375}{0.13375}$$

$$= 0.925$$

Results for various values of k are tabulated on Table 8.2.

TABLE 8.2

$P(n|k \text{ 'heads'})$	k
0.961	2
0.925	3
0.756	5
0.607	6
0.436	7

Two points can be made on the basis of this example:

1. It has been stipulated in this case that the prior probabilities (0.99; 0.01) are arrived at subjectively. It is in such cases that the Bayesian approach provokes most argument and this is centred on the validity and interpretation of subjective probabilities. It is also in these circumstances that the Bayesian approach is most tempting allowing the combination of subjective 'strengths of belief' with empirical evidence. Prior probabilities can of course govern the modified prob-

abilities in a very direct way but this effect is 'diluted' as more and more empirical data is utilized.

2. The probabilities obtained in this case can be interpreted as 'strengths of belief' but not so naturally as relative frequency measures (*see* Section 3.1).

These two points exemplify the main grounds for criticism of the Bayesian approach to problem-solving though the method is currently fashionable and in fact provides a *type* of solution usually inaccessible by other methods.

8.5 DISCRIMINATING POWER OF TEST PROCEDURES

In a number of situations test procedures are proposed which purport to be able to classify individuals or conditions. Aptitude testing and medical diagnostic tests are typical examples. In such cases Bayesian analysis can be informative.

Example. A management consultancy firm offers a sales aptitude testing service. It is claimed that when this test is applied to a group of 'top-class' salesmen 95% pass and 5% fail. When it is applied to a group of average salesmen 10% pass and 90% fail. The company considering use of this service believes that only 1 in 10 of salesmen interviewed has 'top-class' potential and that its current selection procedure results in a mix of 60% 'top-class' salesmen and 40% 'average' salesmen. Would use of the aptitude test at recruitment lead to a higher proportion of top-class salesmen in its sales force?

Using '$T.C$' and 'AV' to indicate the two classes of salesmen and 'pass' and 'fail' to indicate aptitude test performance the above information can be expressed in probability notation thus:

$$P(\text{pass}|T.C) = 0.95, \ P(\text{fail}|T.C) = 0.05$$
$$P(\text{pass}|AV) = 0.10, \ P(\text{fail}|AV) = 0.90$$
$$P(T.C) \quad\ = 0.10, \ P(AV) \quad\ = 0.90$$

The company is interested in the value of $P(T.C.|\text{pass})$. By Bayes' Theorem this is:

$$P(T.C|\text{pass}) = \frac{P(\text{pass}|T.C) \times P(T.C)}{P(\text{pass}|T.C) \times P(T.C) + P(\text{pass}|AV) \times P(AV)}$$
$$= \frac{0.95 \times 0.10}{(0.95 \times 0.10) + (0.10 \times 0.90)} = \frac{0.095}{0.095 + 0.090} = \frac{0.095}{0.185}$$
$$= 0.513$$

The relative frequency interpretation of probability is valid here and hence one can say that employment of the test would result in a sales force containing 51.3% 'top-class' salesmen compared to the present selection procedure result of 60%. The aptitude test is therefore less discriminating than the current selection procedure.

The reader is asked to note that this type of result is not uncommon. Tests with apparently high discriminating power can be shown to perform poorly in practice because the incidence of the desired characteristic in the population to which the test is applied is very low. A crude analogy is that the mesh of a fishing net can only be assessed in relationship to the size of fish to be caught!

8.6 BAYES' THEOREM AND SAMPLING

In Chapters 4 and 5 emphasis was laid on the idea that samples are taken to obtain information and that information costs money and/or time. As Bayes' Theorem is concerned with updating probabilities in the light of information obtained it is interesting to consider the use of the theorem in a sampling situation. The following example has been invented to illustrate the mechanism and argument used and also to illustrate the idea of a *decision rule*.

In many decision-making situations it may not be possible to prescribe the best decision but it may be possible to provide a series of rules of the form:

'If A is observed the best course of action to adopt is X_1.
'If B or C is observed the best course of action to adopt is X_2.
'In all other circumstances the best course of action to adopt is X_3.'

These statements constitute a decision rule in that they prescribe the appropriate course of action conditional upon the phenomena observed at the time of decision. The reflective reader may wish to compare this with the concept of estimator outlined in Chapter 4. In both cases the 'problem solver' is presenting a 'procedure' to be applied to obtain the solution to a problem rather than presenting a 'solution' to the problem.

8.6.1. BAYESIAN DECISION RULES AND SAMPLE SIZE

A company has only one source of partially finished components. These are supplied in batches which may contain either 1% defectives (high quality) or 10% defectives (low quality). Experience reveals that 80% of batches are high quality and 20% are low quality. Acceptance of a low-quality batch involves the company in rectification costs of £200/batch while wrongful rejection of a high-quality batch involves a penalty cost of £60/batch. (Rejection means returning a batch to the supplier for rectification.) It is proposed to base acceptance and rejection of batches on the information obtained by testing a sample of items from each batch. If testing a single component costs £2·50 how should the sampling scheme be set up to minimize expected total cost?

Two questions have to be answered before the sampling scheme can be specified:

How many items should be tested?

How many defectives have to be found to warrant rejection of the batch?

Bayes' Theorem provides an approach to this problem which involves calculating the minimum cost rejection rule for each possible sample size. The notation is self evident and involves typically $P(1d|HQ)$ = Probability of 1 defective given that the batch is high quality. Sample size is designated k.

The courses of action possible and the associated penalties can be shown conveniently in tabular form (Table 8.3).

TABLE 8.3

Penalty costs associated with different actions

	Nature of batch	
Course of action	*HQ*	*LQ*
Accept batch	0	£200
Reject batch	£60	0

$k = 0$ (*no items are tested*). In this case historical experience indicates that if every batch is 'accepted', on 20% of occasions the £200 penalty will be incurred and if every batch is rejected the £60 penalty will be incurred on 80% of occasions.

Thus one can write:

Policy	*Expected cost*
Accept	$0.20 \times £200 = £40$
Reject	$0.80 \times £60 = £48$

Hence for $k = 0$ the better policy is 'Accept'.

$k = 1$ (*1 item tested*). The test result may be either non-defective ($0d$) or defective ($1d$). In either case a sampling cost of £2·50 is necessarily incurred. Suppose the test result is $0d$. By Bayes' Theorem:

$$P(HQ|0d) = \frac{P(0d|HQ) \times P(HQ)}{P(0d|HQ) \times (PHQ) + P(0d|LQ) \times P(LQ)}$$

$$P(0d|HQ) = 0.99; \ P(0d|LQ) = 0.90$$

Hence:

$$P(HQ|0d) = \frac{(0.99 \times 0.80)}{(0.99 \times 0.80) + (0.90 \times 0.20)} = \frac{0.792}{0.972} = 0.815$$

Because $P(LQ|0d) = 1 - P(HQ|0d)$

$$P(LQ|0d) = 0.185$$

The penalty costs associated with each course of action are unchanged but the probabilities of incurring these penalty costs have changed once the item tested is shown to be non-defective. Combining these new probabilities with penalty costs and adding in the sampling cost gives:

Expected cost of accepting batch $= 0.185 \times £200 + £2.50 = £39.5$
Expected cost of rejecting batch $= 0.815 \times £60 + £2.50 = £51.4$

Hence if the test result is $0d$ the better policy is to accept the batch. Now consider the situation if the test result is defective.

$$P(1d|HQ) = 0.01; \quad (1d|LQ) = 0.10$$

By Bayes' Theorem:

$$P(HQ|1d) = \frac{0.01 \times 0.80}{(0.01 \times 0.80) + (0.10 \times 0.20)} = \frac{0.008}{0.028} = 0.286$$

$$P(LQ|1d) = 1.0 - 0.286 = 0.714$$

Expected cost of accepting batch $= 0.714 \times £200 + £2.50 = £145.3$
Expected cost of rejecting batch $= 0.286 \times £60 + £2.50 = £19.66$

Hence if the test result is $1d$ the better policy is to reject the batch.

These results can be presented as in Table 8.4 which leads directly to the *decision rule*.

TABLE 8.4

Expected costs associated with courses of action based on sample results ($k = 1$)

	Test results ($k = 1$)	
Course of action	0d	1d
Accept	£39·50	£145·30
Reject	£51·40	£19·66

If a sample of size 1 is non-defective accept the batch; if defective reject the batch.

This decision rule applies only after the test result is obtained.

The probabilities of getting these test results can be calculated.

$P(0d) = P(0d|HQ) \times P(HQ) + P(0d|LQ) \times P(LQ)$ (a batch must be either HQ or LQ)

Similarly:

$P(1d) = P(1d|HQ) \times P(HQ) + P(1d|LQ) \times P(LQ)$

These expressions are of course the denominators in the appropriate Bayes' Theorem calculations just carried out, therefore:

$$P(0d) = 0 \cdot 972; \; P(1d) = 0 \cdot 028$$

If one takes the lower cost decision, then on 97·2% of occasions one will incur the penalty of £39·5 and on 2·8% of occasions the penalty of £19·66. Combining these costs and probabilities gives:

Expected cost of using the decision rule based on a sample of size 1 =
$0 \cdot 972 \times £39 \cdot 5 + 0 \cdot 028 \times £19 \cdot 66 = £38 \cdot 39 + £0 \cdot 55 = £38 \cdot 94$

This is less than the cost of making a decision in the absence of sample information therefore a sample of size 1 can be justified.

$k = 2$. Sampling cost $= 2 \times £2 \cdot 50 = £5 \cdot 00$

The test results may be $0d$, $1d$, or $2d$ and costs of decisions based on each result are carried out just as for $k = 1$. The only minor complication is that the binomial distribution must be invoked in order to calculate the probable test results. Thus:

$$P(2d|HQ) = {}^{2}C_{2} \times (0 \cdot 01)^{2} = 0 \cdot 0001; \; P(2d|LQ) = {}^{2}C_{2}(0 \cdot 10)^{2} = 0 \cdot 01$$
$$P(1d|HQ) = {}^{2}C_{1} \times (0 \cdot 01 \times 0 \cdot 99) = 0 \cdot 0198; \; P(1d|LQ) = {}^{2}C_{1} \times (0 \cdot 10 \times$$
$$0 \cdot 90) = 0 \cdot 18$$
$$P(0d|HQ) = {}^{2}C_{0} \times (0 \cdot 99)^{2} = 0 \cdot 9801; \; P(0d|LQ) = {}^{2}C_{0} \times (0 \cdot 90)^{2} = 0 \cdot 81$$

Using Bayes' Theorem the revised probabilities for $0d$ are:

$$P(HQ|0d) = \frac{0 \cdot 7841}{0 \cdot 9461} = 0 \cdot 829; \; P(LQ|0d) = 0 \cdot 171$$

Hence:

Expected cost of acceptance $= 0 \cdot 171 \times £200 + £5 \cdot 00 = £39 \cdot 2$
Expected cost of rejection $\quad = 0 \cdot 829 \times £60 \; + £5 \cdot 00 = £54 \cdot 7$

Similarly:

$$P(HQ|1d) = \frac{0 \cdot 01584}{0 \cdot 05184} = 0 \cdot 305 \; P(LQ|1d) = 0 \cdot 695$$

$$P(HQ|2d) = \frac{0 \cdot 00008}{0 \cdot 00208} = 0 \cdot 038 \; P(LQ|2d) = 0 \cdot 962$$

Calculating the expected cost of acceptance and rejection policies gives the values shown in Table 8.5.

The decision rule here is obviously:

If a sample of size 2 has zero defectives accept the batch, otherwise reject it.

TABLE 8.5

**Expected costs associated with courses of action
based on sample results ($k = 2$)**

| | *Test results ($k = 2$)* | | |
Course of action	0d	1d	2d
Accept	£39·20	£144·00	£197·40
Reject	£54·70	£23·30	£7·28

The values $P(0d) = 0·9461$, $P(1d) = 0·0518$, and $P(2d) = 0·0021$ are obtained as before from the denominators of the Bayes expressions and the expected cost of this decision rule can be calculated.

Expected cost of using the decision rule based on a sample of size 2 =
$0·9461 \times £39·2 + 0·0518 \times £23·3 + 0·0021 \times £7·28 = £38·31$

This is a lower cost decision rule than that based on $k = 1$, therefore the costs associated with $k = 3$ must be examined.

Only the salient results will be given from here on. Their derivation is left to the reader as an exercise.

$k = 3$. Sampling cost $= 3 \times £2·50 = £7·50$ (*see* Table 8.6)

TABLE 8.6

| | *Test results ($k = 3$)* | | | |
Course of action	0d	1d	2d	3d
Accept	£39·10	£141·90	£199·10	£206·70
Reject	£58·00	£27·20	£10·00	£7·70

Decision Rule: If a sample of size 3 has zero defectives accept the batch, otherwise reject it.

$P(0d) = 0·9220$; $P(1d) = 0·0716$; $P(2d) = 0·0056$; $P(3d) = 0·0002$

Expected cost of decision rule ($k = 3$) $= £38·1$.

This is less than the cost of the decision rule for $k = 2$, therefore $k = 4$ can be examined.

$k = 4$. Sampling cost $= 4 \times £2·50 = £10$. Results in Table 8.7.

Decision rule: If a sample of size 4 has zero defectives accept the batch, otherwise reject.

$P(0d) = 0·8995$; $P(1d) = 0·0894$; $P(2d) = 0·0102$; $P(3d) = 0·0007$;
$P(4d) = 0·0002$

TABLE 8.7

Test results ($k = 4$)

Course of action	0d	1d	2d	3d	4d
Accept	£39·20	£140·60	£200·80	£209·20	£209·90
Reject	£61·20	£30·80	£12·80	£10·20	£10·00

Expected cost of decision rule ($k = 4$) = *£38·2*.

This is more than the cost of the decision rule for $k = 3$ (though only very slightly more) and consequently $k = 5$ need not be examined as it can now be seen that the additional sampling cost will outweigh the economic benefit of additional information. For the sake of demonstration, however, the results for $k = 5$ are given in Table 8.8.

TABLE 8.8

Test results ($k = 5$)

Course of action	0d	1d	2d	3d	4d	5d
Accept	£39·30	£138·30	£202·10	£211·50	£212·40	£212·50
Reject	£64·50	£34·80	£15·60	£12·80	£12·50	£12·50

$k = 5$. Sampling cost $= 5 \times £2·50 = £12·50$ (*see* Table 8.8)

Decision Rule: If a sample of size 5 has zero defectives accept, otherwise reject.

$P(0d) = 0·8787$; $P(1d) = 0·1040$; $P(2d) = 0·0154$; $P(3d) = 0·0016$; $P(4d) = 0·0001$; $P(5d) = 0·0001$

Expected cost of decision rule ($k = 5$) = £38·4.

The cost of using this decision rule is, as expected, greater than that of using the decision rule for samples of size 4 and no further sample sizes will be considered.

The results of this analysis can be summarized as shown in Table 8.9.

TABLE 8.9

Sample size	0	1	2	3	4	5
Expected cost of decision rule	£40·00	£38·90	£38·30	£38·10	£38·20	£38·40

The information in Table 8.9 can be expressed in words as follows:
For each sample size the decision rule giving the lowest expected cost has been

determined. As sample size increases the more 'reliable' information so obtained reduces the risk of incurring penalty costs but as this information is only obtained at a price, a sample size is reached at which the cost of obtaining additional information is greater than the expected savings which that information can provide. In the example above that sample size is 3.

It may be thought that the expected saving of £1·7/batch is not very impressive. This is of course a consequence of the numerical values used in the example and for other situations the savings may be very substantial. Equally £1·7/batch could only be valued with reference to the number of batches involved and the cost of each batch.

A further point which should be noted is that in this example the decision rule in each case was the same, namely, accept a batch only if the sample taken contains zero defectives. In other situations the rule might vary with sample size and it is possible to imagine situations in which for a sample of size 4 the decision rule would be 'accept batch if sample contains zero defectives', while for a sample of size 5 the decision rule would be 'accept batch if sample contains no more than 1 defective'.

8.7 SUMMARY

Where the value of a probability is specified in relation to certain necessary knowledge it is termed a conditional probability and is written $P(A|B) = x$.

The probability of the joint occurrence of two events A and B where knowledge of the occurrence (or non-occurrence) of one affects the probable occurrence (or non-occurrence) of the other can be calculated using whichever of the equations below is appropriate:

$$P(AB) = P(A|B) \times P(B)$$
$$\text{or } P(AB) = P(B|A) \times P(A)$$

Where the occurrence of an event A is always associated with the occurrence of one of the mutually exclusive and exhaustive events $X_1, X_2, \ldots X_n$ one can write:

$$P(X_i|A) = \frac{P(A|X_i) \times P(X_i)}{P(A|X_1) \times P(X_1) + P(A|X_2) \times P(X_2) + \ldots + P(A|X_n) \times P(X_n)}$$
$$= \frac{P(A|X_i) \times (P|X_i)}{\sum_{i=1}^{n} P(A|X_i) \times P(X_i)}$$

This expression is known as Bayes' Theorem and provides, in a variety of circumstances, a mechanism for updating probabilities in the light of information obtained. In the notation used above the $P(X_i)$s are termed prior likelihoods. Criticism of the use of Bayes' Theorem tends to focus on the origins and interpretation of the prior likelihoods used in specific cases and where subjectively determined probabilities are being updated by experimental information it should be recognized that that choice of prior likelihoods has a major influence on the modified probabilities ultimately obtained. Also careful consideration should be given to whether or not a relative frequency interpretation of that probability is meaningful.

EXERCISES

1. A survey intended to relate television set possession characteristics and car ownership characteristics of a sample of families in a region produces the following table:

| | | Television set | |
Car	No set	Black & white set	Colour set
No car	200	550	50
Car > 2 yrs old	100	600	200
Car ≤ 2 yrs old	20	130	150

What estimates can you make of the following probabilities?

   (a) P (No TV set|Owns car ≤ 2 yrs old)
   (b) P (Colour TV set *and* car ≤ 2 yrs old)
   (c) P (No car|Black and white TV set)
   (d) P (Car owner|TV set)

2. In the example in Section 8.4 the calculations show the effect of 'updating' the prior likelihoods by 'blocks' of information corresponding to 'k' heads being obtained in trials. Show that the same value for $P(n|k$ heads) is obtained if one takes $P(n|(k - 1)$ heads) as a new prior likelihood and updates this by the information corresponding to $1h$ being obtained on the next trial.

3. In the example in Section 8.5 how would you view the proposed test if 1 in 5 of the candidates interviewed had 'top-class' potential?

4. Determine sampling policies for the problem situation posed in Section 8.6.1 if:

   (a) rejection of a high-quality batch involves a penalty cost of £100/batch; and
   (b) testing a single component costs £5.

9. *Index Numbers*

9.1 INTRODUCTION

Most people in all walks of life are interested in how rapidly the cost of living is rising. The Government through the Central Statistical Office (C.S.O.) publishes the General Index of Retail Prices each month and this index is a guide to changes in the cost of living. There are many other facets of the economy whose values change over time and which may be compared at different points in time by means of index numbers. The Monthly Digest of Statistics published by C.S.O., and issued through H.M.S.O., contains a wealth of such information. Successive values of index numbers tend to show up inflationary or deflationary trends in the economy.

In general an index number summarizes a large quantity of information on, say, prices as in the index of retail prices. Index numbers of wage rates summarize wage information for a number of different categories of workers. The index of industrial production summarizes production data for most manufacturing industries. Although the indices mentioned are calculated by or for the C.S.O., there is no reason why an organization should not produce an index number for its own production quantities or other variables.

This chapter gives a brief description of how index numbers are calculated and manipulated. However it is not expected that the reader will wish to calculate his own but the intention is that he should have a better understanding of the basis of the Government statistics. The official index numbers may well be used as variables in further analyses such as multiple regression (Chapter 12).

9.2 CONSTRUCTION OF INDEX NUMBERS

The following discussion takes place in the context of index numbers for prices since these are most commonly used but the methods are equally applicable to construction of index numbers for production quantities, exports, imports or any other variable of interest. The primary function of price index numbers is to compare prices in one year with those in some other year. Technically prices in a *given year* are to be compared with prices in the *base year* which are taken as a standard. Conventionally p_1 refers to the price in the given year and p_0 refers to the price in the base year.

If only one item such as bread is being considered the comparison between years may be made by the calculation of price relatives, i.e. the price in the given year relative to the base year.

Price relative $= (p_1/p_0) \, . \, 100$

e.g., If the price of a loaf was 10p in 1974 and 20p in 1977, the 1977 price relative to 1974 was $(20/10) \, . \, 100 = 200$.

If more than one item or commodity is to be considered to give an overall impression of rising or falling prices it becomes necessary to combine the prices of these items into some form of a weighted average or index number. The most commonly used form is that calculated by the Laspeyres formula:

$$I_1 = \frac{\Sigma \, p_1 q_0}{\Sigma \, p_0 q_0} \, . \, 100$$

where $I_1 =$ index number for the given year

$q_0 =$ weight applied to each price calculated for the base year

Σ to be taken over all the items

The calculation is illustrated using the data of Table 9.1. I_1 is then the index number for 1977 and I_0 would be its value in the base year 1974 and would equal 100.

$$\text{Then } I_1 = \frac{20 \, . \, 100 + 12 \, . \, 400 + 52 \, . \, 50}{10 \, . \, 100 + 6 \, . \, 400 + 29 \, . \, 50} \times 100$$

$$= 193 \cdot 8$$

It may now be stated that prices have risen by $93 \cdot 8 \%$ overall from 1974 to 1977 based on the evidence of these three commodities.

In this example each price is weighted by the quantity consumed of the commodity in the base year. Each term $p_0 q_0$ is the value of the quantity consumed in the base year and the Laspeyres index gives the change in the total value of the base year consumption when valued at given year prices. It is thus a *base year weighted index*. This index is a reasonable measure of the change in prices over a short period of, say, two years, but if the given year is a longer period in time from the base year the weights used tend to become out of date as spending habits change and no longer give a realistic comparison between the two years. In particular if the prices of certain commodities increase considerably, the consumer may well substitute other cheaper commodities for them, e.g. substitution of margarine for butter. The quantity of butter consumed in the given year may then be less than that consumed in the base year while the consumption of margarine may have increased. The net effect is that total values may not have increased as much as that implied by the index which will then have over-estimated the price increases.

TABLE 9·1

	1974		1977	
Commodity	Quantity consumed	Price	Quantity	Price
	q_0	p_0	q_1	p_1
Bread	100 loaves	10p/loaf	150	20p
Milk	400 pints	6p/pint	500	12p
Eggs	50 dozen	29p/dozen	50	52p

This disadvantage may be overcome by using a given year weighted index as calculated by the Paasche formula:

$$I_1 = \frac{\Sigma p_1 q_1}{\Sigma p_0 q_1}$$

This index gives the change in the total value of the given year consumption from the value it would have had in the base year. Using this formula the weights are always up to date. It is applied to the data in Table 9.1.

$$I_1 = \frac{20.150 + 12.500 + 52.50}{10.150 + 6.500 + 29.50} \times 100$$

$$= 195·0$$

From this calculation prices may be said to have risen 95·0% overall. However this formula is equally unrealistic in that it compares hypothetical past quantities with current real quantities rather than vice versa. One suggested way out of the dilemma is to calculate an average index from the two above. Fisher's ideal index number is the geometric mean of the Laspeyres and the Paasche index numbers:

$$I_F = \sqrt{(I_L \cdot I_P)} = \sqrt{\left(\frac{\Sigma p_1 q_0}{\Sigma p_0 q_0} \cdot \frac{\Sigma p_1 q_1}{\Sigma p_0 q_1} \right)}$$

The Paasche formula and hence Fisher's suffer from two more disadvantages. The first is that since each given year's index number is calculated with new weights the only comparisons that can be made are between the given year and the base year and successive years are not directly comparable as they are with the Laspeyres formula. Secondly it is a costly and time-consuming operation to find new weights each year (see below) and the result is that it is convenient to use the Laspeyres formula in most cases. In theory the weights should then be updated fairly frequently.

So far some possible formulae for index numbers have been given and an argument put forward for the use of the Laspeyres formula. It is now necessary to discuss the components of the formula in more detail.

An index number compares values such as prices in a given year with those in a base year or period. The base period is therefore being used as a standard or normal period and should be chosen as a period in which there were no abnormal values. These abnormal values may be difficult to define and a semblance of normality may be achieved by taking average values over, say, three years as base values. However most of the C.S.O. index numbers take a single year as a base year. As a general rule the base period should not be taken as an abnormally 'good' year, e.g. low prices, since all following years will look poor, i.e. have index numbers greater than 100, or conversely taken as an abnormally 'bad' year, e.g. high prices. Of course with price indices any base year is likely to look good due to the effects of inflation. However production output index numbers should have a base year chosen to be roughly in the middle of their possible range. The other factor to bear in mind in the choice of a base year is that it should not be too far in the past since the base year weights will become obsolete.

Index numbers measure change in values and it is necessary to define precisely what change is to be measured. Thus it is necessary to define the prices and weights to be used precisely. For example what price should be taken for the index of retail prices? Is it the manufacturers' recommended price, or the price on the label in the shop, or the price paid by the customer after some form of discount? Is the discount the same for each customer? The index of retail prices is supposed to reflect changes in prices nationally but is the price paid in London the same as that paid in Glasgow? This gives rise to sampling problems.

Finally the weights used must be relevant to the questions being asked of the index number. Thus if an index is designed to show the change in price of the housewife's shopping basket from 1974 to 1977 it is necessary to know whether she buys the same quantity of each commodity in 1977 as she did in 1974. If not the weights applied to each commodity should theoretically be updated for 1977. However it would be necessary to carry out each year a survey of consumer buying patterns to keep the weights up to date. This could be a very costly exercise. In fact the weights are continually updated for the index of retail prices but other index numbers use only base year weights which are updated every ten to fifteen years. Generally each time the base year is changed the weights are updated.

9.3 MANIPULATION OF INDEX NUMBERS

This section discusses how an index number series may be recalculated on a new base year, how new weights may be introduced into the series, and how the series may be used to calculate deflated or 'real' prices.

9.3.1 CHANGING THE BASE

The base of an index number series is changed by taking proportions as illustrated in Table 9.2. Index *A* has 1971 as a base year and Index *B* has 1976 as a

TABLE 9.2

Base change

Year	Index A	Index B
1971	100	66·7
1972	110	73·3
1973	120	80·0
1974	130	86·7
1975	140	93·3
1976	150	100

base year. To convert Index A to Index B each Index A value was divided by 150. It can be seen that the numbers for each year are in the same proportion for both Index A and Index B.

9.3.2 CHAIN INDEX NUMBERS

A chain index number takes the previous year as its base and compares prices in the given year with those in the previous year.

Example. If $I_{75,74}$ is the chain index for 1975 based on 1974 prices then:

$$I_{75,74} = \frac{\Sigma p_{75}q_{74}}{\Sigma p_{74}q_{74}} \text{ using Laspeyres formula}$$

Suppose 1974 weights are to be used as base weights for each chain index, then:

$$I_{76,75} = \frac{\Sigma p_{76}q_{74}}{\Sigma p_{75}q_{74}}$$

Multiplication of these two chain index numbers gives the normal Laspeyres formula for the index in 1976, based on 1974 weights and prices.

$$I_{76} = I_{76,75} \cdot I_{75,74} \qquad\qquad [9.1]$$
$$= \frac{\Sigma p_{76}q_{74}}{\Sigma p_{75}q_{74}} \cdot \frac{\Sigma p_{75}q_{74}}{\Sigma p_{74}q_{74}}$$
$$= \frac{\Sigma p_{76}q_{74}}{\Sigma p_{74}q_{74}}$$

where I_{76} is the normal index for 1976 based on 1974.

Use of Equation [9.1] gives another method for calculating the price index in the next given year.

Example. If 1974 is the base year and I_{76} is already calculated, then in 1977 it is only necessary to calculate the chain index on 1976, $I_{77,76}$, and multiply that by I_{76} to give I_{77}.

$$I_{77} = I_{76} \cdot I_{77,76}$$

The following numerical example uses the data in Table 9.3 and takes 1974 as the base year.

$$I_{75,74} = I_{75} = \frac{\Sigma p_{75}q_{74}}{\Sigma p_{74}q_{74}} = \frac{7,660}{5,850} = 1\cdot309$$

$$I_{76,75} = \frac{\Sigma p_{76}q_{74}}{\Sigma p_{75}q_{74}} = \frac{8,890}{7,660} = 1\cdot161$$

$$I_{76} = \frac{\Sigma p_{76}q_{74}}{\Sigma p_{74}q_{74}} = \frac{8,890}{5,850} = 1\cdot520$$

or $I_{76} = I_{75} \cdot I_{76,75} = 1\cdot309 \cdot 1\cdot161 = 1\cdot520$.

A single chain index number, e.g. $I_{76,75} = 1\cdot161$, is sometimes known as a *link relative*.

TABLE 9.3

Chain index numbers

Commodity	1974 q_{74}	1974 p_{74}	1975 p_{75}	1976 p_{76}	$p_{74}q_{74}$	$p_{75}q_{74}$	$p_{76}q_{74}$
Bread	140	10	14	16	1,400	1,960	2,240
Milk	500	6	7·5	9	3,000	3,750	4,500
Eggs	50	29	39	43	1,450	1,950	2,150
					5,850	7,660	8,890

9.3.3 SPLICING OVERLAPPING SERIES OF INDEX NUMBERS

Suppose index A has a base of 1972 and that in 1974 it becomes necessary to alter the weights used, thus producing a new index, B, based on 1974. However it is not very meaningful to have an index series covering only three years such as A, but continuity would be maintained if the new series B could be expressed in terms of the series A. The process of combining such overlapping series is known as splicing. The process is really one of taking proportions using a chain index and it is illustrated using the data in Table 9.4.

TABLE 9.4

Splicing index numbers

Year	Index A Σpq_{66}	Index B Σpq_{68}
1972	240	
1973	200	
1974	180	200
1975		180
1976		160

$$Series\ A:\ I_{72} = \frac{240}{240} \cdot 100 = 100$$

$$I_{73} = \frac{200}{240} \cdot 100 = 83 \cdot 3$$

$$I_{74} = \frac{180}{240} \cdot 100 = 75$$

$$Series\ B:\ I'_{74} = \frac{200}{200} \cdot 100 = 100$$

$$I'_{75} = \frac{180}{200} \cdot 100 = 90$$

$$I'_{76} = \frac{160}{200} \cdot 100 = 80$$

The chain index numbers for series B are:

$$I'_{75,74} = \frac{180}{200} = 0 \cdot 9$$

$$I'_{76,75} = \frac{160}{180} = 0 \cdot 89$$

One can expect the ratio 1975 to 1974 to be the same for both index A and index B, i.e.

$$\frac{I_{75}}{I_{74}} = \frac{I'_{75}}{I'_{74}}$$

$$I_{75} = \frac{I'_{75}}{I'_{74}} \cdot I_{74} = \frac{90}{100} \cdot 75 = 67 \cdot 5$$

It can be seen that I'_{75}/I'_{74} is the definition of the chain index $I'_{75,74}$ and therefore the formula for calculating I_{75} may be rewritten as:

$$I_{75} = I_{74} \cdot I'_{75,74}$$
$$= 75 \times 0 \cdot 9 = 67 \cdot 5$$

In general the next value in series A, I_{k+1}, may be obtained by multiplying the previous value, I_k, by the equivalent chain index for series B, $I'_{k+1,k}$

$$I_{k+1} = I_k \cdot I'_{k+1,k}$$

The index series B came into being because the weights were changed in 1974. It would of course be possible to change the weights every year and using the chain index technique relate that year back to the original base series A. This is the method used in calculating the index of retail prices.

9.3.4 DEFLATING PRICES AND INCOMES

Indicators of inflation are rising prices and incomes. The question sometimes asked is: by how much has real income increased in, for example, the past two years? It may be rephrased as: if the effect of inflation is discounted, by how much has income increased? It may be answered by deflating the income figures by dividing by the retail price index. Prices of individual commodities may be deflated in the same manner, thus showing the increase in real price.

TABLE 9.5

Deflating income

Year	Income	Price index	Real income
1974	£5,800	100	£5,800
1976	£7,000	157	£4,460

Example. Suppose that the income column in Table 9.5 shows the incomes for a sales representative in 1974 and 1976. The base of the index of retail prices has been taken as 1974 and the value for 1976 is 157. Real income may be calculated by dividing actual income by the price index.

$$1974 \text{ real income} = \frac{£5,800}{1\cdot00} = £5,800$$

$$1976 \text{ real income} = \frac{£7,000}{1\cdot57} = £4,460$$

It may now be said that the salesman's real income has decreased by £1,340 over the two years.

9.4 THE INDEX OF RETAIL PRICES

This index is often thought of as a cost of living index but in fact it is only an index of price changes. A cost of living index would need to measure changes in the amounts and kinds of goods that people buy. The retail price index measures changes in cost of a large 'shopping basket' of goods. It covers the following groups of items: food, alcoholic drink, tobacco, housing, fuel and light, durable household goods, clothing and footwear, transport and vehicles, miscellaneous goods, services. Each group is broken down into sections such as coal and coke, gas, and electricity in the fuel and light group. Each group and section is given a weight such that the total weight for all items is 1,000. The weights may be looked upon as the proportions of 1,000 pence spent on each section by an average household.

The base period is currently taken as January 1974 and the index is recalculated each month and published by section, group, and total in the *Monthly Digest of Statistics*. The weights are updated in January of each year but continuity of the index series is maintained by the use of the chain index technique. The new weights are derived from the results of the continuous Family Expenditure Survey for the three years ended in the June preceding the date of revision. Further information may be obtained from the H.M.S.O. publication: *Method of Construction and Calculation of the Index of Retail Prices*, 1967.

9.5 SUMMARY

Index numbers are used to summarize information on prices, industrial output, international trade, etc., such that comparisons may be made from year to year. The most commonly used formula is that of Laspeyres:

$$I_1 = \frac{\Sigma\, p_1 q_0}{\Sigma\, p_0 q_0}$$

This formula enables successive years to be compared directly. In some cases it is necessary to update the weights, q_0, fairly frequently but continuity of the base series may be maintained by the splicing technique. Indices may be used to deflate prices and incomes and the most commonly used index in this context is the index of retail prices. Yamane [1] has a chapter on the theory and practice of index numbers which includes further information on the index of retail prices and the index of industrial production. Thirkettle [2] has chapters on published statistics which tend to be in index number form.

EXERCISES

1. A firm wishes to follow the relative movement of the prices of the raw materials which it uses. It decides to construct a simple index number starting from January 1978 and based on average prices from 1976. From the following data calculate an index for the month of January 1978:

Raw Materials	Price (£ per tonne) Average, 1976	January 1978	Weight
A	16	19	5
B	24	25	1
C	13	18	3
D	8	9	6
E	12	14	4
F	4	8	3

(From IM Part 2)

2. The following data represent the price per unit of four commodities *A, B, C,* and *D* in 1972, 1973, 1974, and 1975.

Commodity	Unit	Quantity used 1972	Prices (pence) in 1972	1973	1974	1975
A	500 grams	3	10	12½	11	12½
B	box	4	12½	15	14	15
C	dozen	2	15	12½	14	16
D	500 grams	1	5	4	6	7½

Calculate an index number for each year, using 1972 as base year.

(IOS, Part 1)

3. A large cake manufacturer seeks the Price Commission's approval for a price increase of his most expensive cake on the basis of increased production costs. Suppose the Board sets a minimum increase of 15% before a price change can be approved. Calculate an index number from the costs and quantities given below to determine whether the claim would be successful.

Ingredient	Cost of ingredient (p) 1976	1977	Quantity used per cake in 1976
A	20·0	40·0	2
B	15·5	14·0	3
C	26·5	31·5	1
D	20·5	22·5	5
E	18·5	16·0	4

How would the decision be affected if the manufacturer changed his ingredients to 3 units of ingredient *D* and 5 units of ingredient *E* in 1977?

(IOS, Part 1)

4. Using the data given in the following table, calculate an Index of Real Wages over the period 1957–62.

Year	Indices of basic wage rates (Jan. 1956 = 100)	Indices of retail prices (Jan. 1956 = 100)
1957	110·7	105·8
1958	113·4	109·0
1959	116·8	109·6
1960	119·9	110·7
1961	125·0	114·5
1962	129·3	117·5

(From IM, Part 2)

5. Calculate as an arithmetic mean, correct to the nearest whole number, a cost of living index from the following table of price relatives and weights:

	Price relatives	Weights
Food	122	35
Rent	101	9
Clothing	118	10
Fuel	115	7
Miscellaneous	108	39

(*From IM, Part 2*)

[*Hint:* The weights may be taken as value weights $p_0 q_0$.]

6. The following data refer to exports of fertilizer in 1968 and 1978:

	Quantity ('000 tonnes)		Values (£'000)	
	1978	*1968*	*1978*	*1978*
Nitrogenous	63·1	24·2	2,672	389
Other	736·3	459·1	7,929	1,833
Totals	799·4	483·3	10,601	2,222

(*a*) Find the average value for each type of export in each year. Revalue 1978 quantities at 1968 values and 1968 quantities at 1978 values.

(*b*) Calculate index numbers of average values and of volume (i) for 1978 on 1968 as base; (ii) for 1968 on 1978 as base.

(*IOS, Part 1*)

7. A price index based on four commodities uses 1970 weights as a base. The value of the index in 1974 is 117 and in 1975 it is 136. The weights are updated in 1975 as shown below and prices are given for 1976 and 1977.

Commodity	Weights		Prices	
	1975	*1975*	*1976*	*1977*
A	40	35	40	47
B	15	11	13	17
C	29	55	60	62
D	14	40	45	51

Calculate the indices for 1976 and 1977 using 1970 base weights.

10. *Relationships Between Factors*

10.1 INTRODUCTION

When comparing sets of figures, one often feels that a relationship should exist between one set and another, i.e. between one factor or variable and another. One of the simplest forms of such relationships is discussed in this chapter. Its extension to more complex forms is shown in principle.

Consider the case of a brand manager of a luxury convenience food product. He knows that sales are strongly affected by the level of advertising and, in addition, he believes that sales are affected by the quality of the distribution service offered to retail outlets. The level of advertising is measured by the quarterly 'above the line' advertising appropriation and the quality of distribution may be measured by the level of unfulfilled orders in each quarter. After two years operations, he has collected data as shown in Table 10.1 and he would now like to estimate the effects of advertising and distribution on sales turnover.

TABLE 10.1

Quarter	Sales turnover £10,000s	Advertising expenditure £1,000s	Unfulfilled orders 100s cases
1	25	11	2
2	25	12	4
3	15	9	4
4	8	3	16
5	13	5	14
6	10	5	13
7	12	6	12
8	20	9	8

First he could look at the data graphically by plotting a time series for each of the three variables (Figure 10.1). However these graphs give little quantitative information. They show qualitatively that advertising expenditure is in step with sales turnover, i.e. perhaps strongly related, and that unfulfilled orders are out of step with turnover, i.e. perhaps strongly inversely related. In general such graphs are useful for obtaining a subjective 'feel' for data but not for obtaining quantitative relationships.

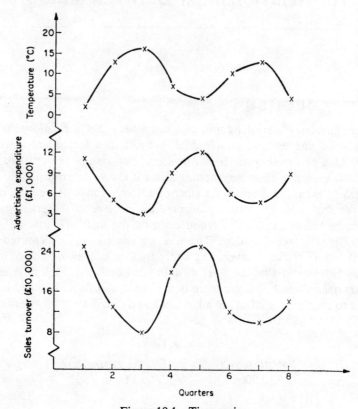

Figure 10.1   Time series

10.2 SCATTER DIAGRAMS

A more useful graphical representation of the relationships may be obtained by plotting one variable against another, e.g. turnover against advertising expenditure as shown in Figure 10.2.

It will be noted that the plotted points are scattered over the diagram and hence the graph is known as a *scatter diagram*. The choice of axes is arbitrary but, by convention, the variable that is to be estimated (known as the dependent variable) is put on the vertical *y*-axis and the other (independent) variable on the horizontal *x*-axis.

The points in Figure 10.2 lie approximately on a straight line and one might hypothesize that there is a linear relationship between turnover and advertising expenditure. This relationship, of course, ignores the effect of distribution. It is possible to obtain a crude estimate of this relationship by drawing a straight line by eye through the points. Although this method may produce a reasonable estimate it gives no indication as to the 'strength' of the relationship nor to the size of the possible error likely to be incurred when using the relationship for predictive purposes. It is therefore necessary to obtain a more objective estimate of the relationship by mathematical means.

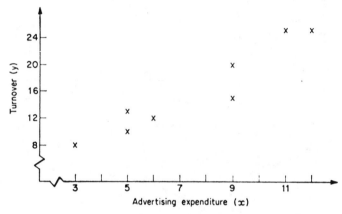

Figure 10.2   Scatter diagram

The points on a scatter diagram do not always lie near a straight line but may produce a 'curvy' pattern. The relationship is then said to be curvilinear and may also be estimated mathematically.

Example 10.1. Draw scatter diagrams for the following sets of paired data suitably choosing the x and y axis.

(a) No. of calls made/representative   7   6   8   6   1   2
    Sales Turnover (£100s)   11   10   14   12   8   9
(b) Market share %   46   7   35   52   45   17   40   48   26   30
    Advertising expendi-
    ture (£1,000s)   45   5   15   65   33   2   26   60   6   20

Solution:

(a) It seems reasonable to endeavour to predict sales turnover from the number of calls made per representative and therefore the former is made the dependent variable y and the latter the independent variable x. (*See* Figure 10.3.)

A curve could be fitted by eye and might be assumed to be linear with an equation of the type:

$$Y = a + bx \quad (\textit{see Section 10.3})$$

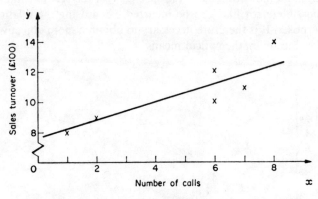

Figure 10.3

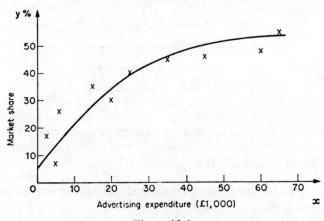

Figure 10.4

(*b*) One might argue that market share depends on the level of advertising or alternatively that the future level of advertising should depend on the current market share. The former argument is chosen in this example and thus market share (*y*) is to be predicted from a given advertising expenditure (*x*). (*See* Figure 10.4.)

The curve drawn in by eye might have an equation of the type:

$$Y = b(1 - e^{-ax})$$

where *a* and *b* are positive constants.

10.3 FITTING A STRAIGHT LINE

As mentioned, Figure 10.2 indicates that there may be a straight line or linear relationship between turnover and advertising expenditure. This section is concerned with objectively fitting a 'best' straight line to the data points.

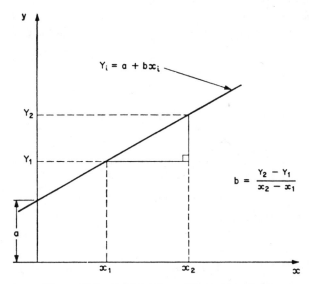

Figure 10.5   Straight line with positive slope

The equation for a straight-line or linear relationship is:

$$Y_i = a + bx_i$$

where a and b are constants.

    a represents the intercept on the y-axis (i.e. the value of Y when $x = 0$)
    b represents the slope or gradient of the line (i.e. the increase in Y for unit increase in x) (*see* Figure 10.5)
    $x_i = i^{th}$ observed value of x (e.g. x_3 means the x value for 3rd quarter)
    $y_i = i^{th}$ observed value of y
    $Y_i =$ an estimate or predicted value of y based on i^{th} value of x

The constants a and b may be positive or negative. Negative slope, b, implies a graph of the form of Figure 10.6, since there is a decrease in Y for unit increase in x. In the above example the expected relationship is:

Turnover $= a + b$ (Advertising Expenditure)

In fitting a line to data, one is locating the line on the graph such that it will pass down the middle of the 'path' formed by the data points. Its location is fixed by calculating values for a and b which in turn specify the relationship between x and Y.

It is reasonable to make the assumption that all the variability in the data is contained in the y-values and that there is no error in the measurement of the x-values. For example expenditure on advertising is specified or fixed by marketing for a given period and a certain level of sales results. If the same advertising expenditure was fixed for another period, it is likely that a different level of

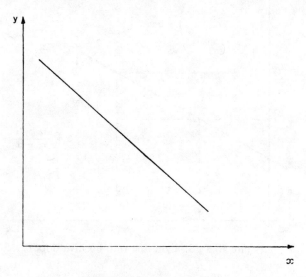

Figure 10.6   Negative slope

sales would result. Thus it would be possible to obtain a distribution of sales levels for one fixed level of advertising expenditure. It is therefore assumed that y-values are observations corresponding to specified levels of x-values.

The criterion for a line to be a good fit is that it should as nearly as possible pass through the means of the y-distributions for each of the x-values. This criterion may be restated as: 'the line of best fit should be such as to minimize the sum of the distances of each y-value from its equivalent point on the line, i.e. for its fixed x-value'. These distances are in fact the vertical distances from the points to the line and are the differences between observed values of y, y_i, and their estimates, Y_i, for fixed values of x, x_i. (*See* Figure 10.7).

Thus the vertical distance, V_i, is given by:

$$V_i = y_i - Y_i \text{ for } x = x_i$$
e.g. $V_3 = y_3 - Y_3$ for $x = x_3$ in Figure 10.7.

The criterion is then that the line should be so chosen as to minimize the sum of the vertical distances from the data points on to the line, i.e. $V_1 + V_2 + V_3 + V_4$ to be a minimum. If the distances are said to be positive for points above the line and negative for points below the line then, on average, the sum of the distances over all points equals zero and the criterion is not very helpful, because a large number of lines satisfying this criterion can be drawn through any set of points. However, if the distances are squared, the figures resulting are all positive and an equivalent criterion is that the line chosen

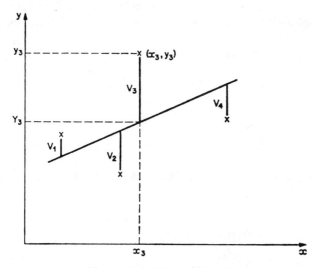

Figure 10.7   Line of best fit

should minimize the sum of the squared distances. This is known as the *Least Squares Criterion*. The line so calculated is known as the *regression line of y on x* (i.e. *y* based on *x*-values). Equally well the line can be calculated so as to minimize the sum of the horizontal distances squared and in this case is known as the *regression line of x on y*. The line *x* on *y* uses the assumption that the variability in the data is contained in the *x*-values and that the *y*-values are fixed. In general, the line *y* on *x* is different to the line *x* on *y* and they are coincident only if the points all lie exactly on a straight line. Providing the variable to be predicted is designated as *y* and the other variable is designated as *x*, then it is only necessary to calculate the line *y* on *x*.

Looking at the situation more statistically the true or functional relationship between *x* and *y* is said to be of the form:

$$Y_i = \hat{a} + \hat{b}x_i$$

But the relationship does not describe the observed values perfectly because of upsetting environmental factors and the relation between observed values of y and x is of the form:

$$y_i = \hat{a} + \hat{b}x_i + \varepsilon_i$$

where ε_i (Greek epsilon) is a 'noise' or 'error' term.

ε_i thus accounts for those observed values of y_i not lying on $(\hat{a} + \hat{b}x_i)$. It is necessary to find values of a and b which 'best' represent the relationship or estimate it so that y may then be estimated for a given value of x. a and b are estimates of the true values $\hat{a}$ and $\hat{b}$. The interpretation of 'best' is that the values of a and b should be such as to minimize the sum of squared errors, i.e. finding the line of best fit.

If S is the sum of squared errors and there are n observations, then:

$$S = \sum_{i=1}^{n} \varepsilon_i^2 = \sum_{i=1}^{n} (y_i - Y_i)^2$$

$$= \sum_{i=1}^{n} (y_i - a - bx_i)^2$$

and is required to be a minimum. It can be shown that for S to be a minimum a and b take values given by:

$$a = \frac{1}{n}(\Sigma y - b \Sigma x) = \bar{y} - b\bar{x}$$

$$b = \frac{n \Sigma xy - \Sigma x \Sigma y}{n \Sigma x^2 - (\Sigma x)^2}$$

where: $\bar{y}$ = mean of y values
$\bar{x}$ = mean of x values

These estimated values of the constants are now substituted in the basic equation from which a value of y can be estimated for a given value of x. It will be noted that the following summations are required from data:

$$\Sigma y, \ \Sigma x, \ \Sigma x^2, \ \Sigma xy$$

In fact Σy^2 is also calculated as it is required to test the 'goodness' of the fit.

It should also be noted that the estimated line $Y_i = a + bx_i$ passes through the point $(\bar{x}, \bar{y})$, i.e. the co-ordinates of the point corresponding to the means of the x and y-values; b is sometimes known as the *regression coefficient*.

TABLE 10.2

Quarter	Turnover	Advertising expenditure			
	y	x	y^2	x^2	xy
1	25	11	625	121	275
2	25	12	625	144	300
3	15	9	225	81	135
4	8	3	64	9	24
5	13	5	169	25	65
6	10	5	100	25	50
7	12	6	144	36	72
8	20	9	400	81	180
	128	60	2,352	522	1,101

In Table 10.2

y = Turnover in £10,000

x = Advertising expenditure in £1,000

Thus $n = 8$, $\Sigma y = 128$, $\Sigma x = 60$, $\Sigma y^2 = 2,352$, $\Sigma x^2 = 522$, $\Sigma xy = 1,101$

$$b = \frac{8 \cdot 1101 - 60 \cdot 128}{8 \cdot 522 - 60 \cdot 60} = \frac{1,128}{576} = 1 \cdot 958$$

$$a = \frac{128}{8} - 1 \cdot 958 \cdot \frac{60}{8} = 1 \cdot 3$$

i.e. $Y_i = 1 \cdot 3 + 1 \cdot 958\, x_i$

But y_i or $Y_i = \dfrac{\text{Turnover}}{10,000}$

and $x_i = \dfrac{\text{Advertising expenditure}}{1,000}$

and thus $\dfrac{\text{Turnover}}{10,000} = 1 \cdot 3 + 1 \cdot 958\, \dfrac{\text{Advertising}}{1,000}$

or Turnover $= 13,000 + 19 \cdot 58$ (Advertising)

Using this equation the brand manager would estimate his turnover at £91,320 for an advertising appropriation of £4,000.

In using this forecasting equation, he is assuming that distribution is just one of a number of unknown factors which cause the noise term, ε_i, and he ignores these for prediction purposes.

Example 10.2. Calculate a linear estimating equation for sales turnover from the data of Example 10.1(*a*).

TABLE 10.3

Sales turnover	No. of calls			
y	x	y^2	x^2	xy
11	7	121	49	77
10	6	100	36	60
14	8	196	64	112
12	6	144	36	72
8	1	64	1	8
9	2	81	4	18
64	30	706	190	347

From Table 10.3:

$$n = 6$$

$$b = \frac{n \, \Sigma \, xy - \Sigma \, x \, \Sigma \, y}{n \, \Sigma \, x^2 - (\Sigma \, x)^2} = \frac{6 \cdot 347 - 30 \cdot 64}{6 \cdot 190 - 30 \cdot 30}$$

$$= \frac{27}{40} = 0 \cdot 675$$

$$a = \bar{y} - b\bar{x} = \frac{64}{6} - \frac{27}{40} \cdot \frac{30}{6}$$

$$= \frac{175}{24} = 7 \cdot 292$$

$$\therefore Y_i = 7 \cdot 292 + 0 \cdot 675 \cdot x_i$$

But y or Y is sales turnover in units of £100

i.e. $Y = \dfrac{\text{Sales turnover}}{100}$

$\therefore \dfrac{\text{Sales turnover}}{100} = 7 \cdot 292 + 0 \cdot 675 \, (\text{No. of calls})$

$\therefore \text{Sales turnover} = 729 \cdot 2 + 67 \cdot 5 \, (\text{No. of calls})$

Example 10.3. Calculate a linear estimating equation for percentage increase in sales from:

% increase in sales	6	8	7	9	5	7	6	8
% increase in salesmen's commission rates (10%)	3	4	5	6	3	4	5	6

Let $y = \%$ increase in sales

$x = \%$ increase in salesmen's commission rates in 10% units

TABLE 10.4

y	x	y^2	x^2	xy
6	3	36	9	18
8	4	64	16	32
7	5	49	25	35
9	6	81	36	54
5	3	25	9	15
7	4	49	16	28
6	5	36	25	30
8	6	64	36	48
56	36	404	172	260

From Table 10.4:

$n = 8$

$$b = \frac{8 \cdot 260 - 36 \cdot 56}{8 \cdot 172 - 36 \cdot 36} = \frac{4}{5} = 0\cdot8$$

$$a = 7 - \frac{4}{5} \cdot \frac{36}{8} = \frac{17}{5} = 3\cdot4$$

$$Y_i = 3\cdot4 + 0\cdot8 \, x_i$$

But $x = \dfrac{\% \text{ increase in commission rates}}{10}$

$$\therefore \% \text{ increase in sales} = 3\cdot4 + 0\cdot8 \, \frac{(\% \text{ increase in commission rates})}{10}$$

$$= 3\cdot4 + 0\cdot08 \, (\% \text{ increase in commission rates})$$

10.4 CORRELATION

Although the quantitative form of the relationship between x and y has been found, it is necessary to establish its 'strength'. One therefore calculates the (product moment) correlation coefficient by:

$$\text{Correlation coefficient} = r = \frac{n \, \Sigma \, xy - \Sigma \, x \, \Sigma \, y}{\sqrt{[(n \, \Sigma \, x^2 - (\Sigma \, x)^2) \, (n \, \Sigma \, y^2 - (\Sigma \, y)^2)]}}$$

r is a measure of how closely the points are scattered about a straight line. Values of r lie between -1 and 1 inclusive and when r equals ± 1, y is perfectly correlated to x, i.e. all the points lie on the straight line and this is assumed to be evidence for a perfect functional relationship between x and y. If $r = 0$, there is said to be no correlation and therefore no evidence for a relationship existing between x and y. The sign of r depends on the slope of the line, i.e. positive b gives positive r, negative b gives negative r.

In the turnover example the correlation coefficient is given by:

$$r = \frac{8 \cdot 1101 - 60 \cdot 128}{\sqrt{(8 \cdot 522 - 60^2)(8 \cdot 2352 - 128^2)}}$$

$$= 0 \cdot 953$$

If r lies somewhere between 0 and ± 1, then the estimated relationship is something between non-existent and perfect. It is then necessary to test the significance of, or the weight that can be attached to the r-value. Using a small number of observations it would be possible to obtain a high value of r by chance alone whereas a larger sample size might have given an insignificant r-value, e.g. one would obtain $r = \pm 1$ when fitting a line to only 2 observations, whether or not these 2 observations are representative of the population. Increasing the sample size increases the likelihood of identifying the true functional relationship and therefore reduces the effects of chance on the size of the correlation coefficient. Thus one would expect the significance of r to depend on the sample size used. The significance of r may be tested by a t-test and the form of the argument is:

Null hypothesis:

The value obtained is due purely to chance and there is no relationship between the variables. If the hypothesis were true one would *expect* the correlation coefficient to be zero. To test the hypothesis one tests whether or not the observed value is significantly different from zero.

Significance level:

1% (say).

Calculation of probability:

Calculate $t = r\sqrt{\dfrac{n-2}{1-r^2}}$ since it may be shown to be distributed like a t-distribution with $(n-2)$ degrees of freedom if the true correlation is zero (i.e. the null hypothesis) and if the data is a sample from a normal population (an assumption made in practice).

This t value is compared with that in the $(n-2)$ row of the 't' table (page 237) at the column corresponding to the chosen significance level. If the value observed is greater than the tabulated value the null hypothesis is rejected. If less the null hypothesis is accepted.

In the above example:

$$r = 0 \cdot 953, \ n = 8$$

Hence:

$$t = 0 \cdot 953 \sqrt{\frac{6}{1 - (0 \cdot 953)^2}}$$

$$= 7 \cdot 7$$

The entry in the '*t*' table in row '6' and 1% column is 3·707, i.e. values greater than this have less than 1% probability of occurring by chance alone. Hence the null hypothesis is rejected and it is concluded that the evidence *does* suggest a real association between turnover and advertising expenditure.

Note 1. In the case of correlation coefficients, tables of significance (in terms of significance levels and number of pairs of values) are available in many statistics textbooks. Thus the 'significance' may be determined directly without reference to '*t*' values.

Note 2. The square of the correlation coefficient, r^2, is a measure of how much of the variance in the *y* values is explained by the linear relationship with *x*. In the above example 90% of the variance in the turnover values is explained by the linear relationship with advertising expenditure.

A significant correlation coefficient does not necessarily indicate a causal relationship. Thus a high level of advertising may not be the cause of a high level of turnover although likely in this example. However the high turnover level might be due to a low level of unfulfilled orders. Many examples of ridiculous superficial correlation may be quoted, e.g. a significant correlation between birth rate and the number of ordinations of priests does not necessarily indicate a causal relationship between them. The moral to be drawn is that there must be some further knowledge of the mechanism linking two variables before one can arrive at a conclusion of causality.

If *A* is known to cause *B* there is likely to be a significant correlation between *A* and *B*. There will also be a significant correlation between *A* and *C* if *A* causes *C*. In addition there is quite likely to be a significant correlation between *B* and *C* but one cannot draw the conclusion that *B* causes *C* or vice versa.

For example, hot weather (*A*) may cause ice-cream sales (*B*) to increase but cause soup sales (*C*) to decrease. There is likely to be a significant positive correlation between temperature and ice-cream sales and a significant negative correlation between temperature and soup sales. There is also likely to be a significant negative correlation between ice-cream sales and soup sales but one would be ill advised to predict ice-cream sales from soup sales or vice versa.

Example 10.4. Calculate the correlation coefficient and test it for significance using the data of:

(*a*) Example 10.2
(*b*) Example 10.3

$$(a)\ r = \frac{n \Sigma xy - \Sigma x \Sigma y}{\sqrt{[(n \Sigma x^2 - (\Sigma x)^2)(n \Sigma y^2 - (\Sigma y)^2)]}}$$

$$= \frac{6 . 347 - 30 . 64}{\sqrt{[(6 . 190 - 30^2)(6 . 706 - 64^2)]}}$$

$$= 0·884$$

$n = 6$

$$t = r \sqrt{\frac{n - 2}{1 - r^2}}$$

$$= 0 \cdot 884 \sqrt{\frac{4}{1 - 0 \cdot 884^2}} = 3 \cdot 78$$

The value of t from tables in row 4 (i.e. 4 degrees of freedom) and column 5% is 2·776. The value calculated is greater than 2·776 and would have a probability of less than 5% of occurring due to chance alone. Hence the null hypothesis is rejected at the 5% level and it is concluded that there is evidence to suggest a real relationship between sales turnover and number of calls made.

$$(b) \ r = \frac{8 \cdot 260 - 36 \cdot 56}{\sqrt{[(8 \cdot 172 - 36^2)(8 \cdot 404 - 56^2)]}}$$

$$= 0.73$$

$$n = 8$$

$$t = 0 \cdot 73 \sqrt{\frac{6}{1 - 0 \cdot 73^2}} = 2 \cdot 62$$

The tabulated value of t using 6 degrees of freedom and the 5% column is 2·447 which is less than that calculated above. Again the null hypothesis is discredited at the 5% level of probability and it is concluded that the data does suggest a relationship between % increase in sales and % increase in salesmen's commission rates.

10.5 RANK CORRELATION

On some occasions the scatter diagram will not indicate high correlation but only indicate a vague linear relationship. In these cases it is worth while testing for significant correlation before carrying out the somewhat lengthy process of fitting a line. If the correlation turns out to be insignificant, then the fitted line will be of little value for predictive purposes and one should then look for some other factor to correlate with the one it is desired to predict. Unfortunately the calculation of the correlation coefficient, r, is also a lengthy process.

Spearman's rank correlation coefficient, ρ, (Greek rho), is a measure of the agreement between two ranked variables. Spearman's ρ lies between -1 and $+1$ and values of near ± 1 indicate a 'strong' relation between the two variables. In this case the relationship does not have to be linear and may take other forms. Thus the rank correlation coefficient has wider uses than the product moment correlation coefficient for ranked variables.

If the original values of the variables are ranked (as shown below) and since Spearman's ρ takes values near ± 1 for linear relationships, it may be used as an approximation for r. It has the advantage of being simple and quick to calculate. The method of calculation is as follows:

1. Rank separately the values of the variables from 1 upwards, preferably in the same direction, e.g. lowest to highest or vice versa.

Let S_i = rank of observation y_i

$\quad R_i$ = ,, ,, ,, x_i

2. Calculate the difference between each pair of rankings.

Let d_i = difference between i^{th} pair of rankings

$\quad\quad = S_i - R_i$

3. Square and sum the differences, d_i, i.e.

$$\sum_{i=1}^{n} d_i^2 = \sum_{i=1}^{n} (S_i - R_i)^2$$

where n = number of paired observations

4. Calculate ρ from:

$$\rho = 1 - \frac{6 \sum_{i=1}^{n} d_i^2}{n(n^2 - 1)}$$

Taking the turnover example, the variables may be ranked lowest to highest as in Table 10.5.

TABLE 10.5

Quarter	y	x	Rank S	Rank R	d	d^2
1	25	11	7·5	7	0·5	0·25
2	25	12	7·5	8	−0·5	0·25
3	15	9	5	5·5	−0·5	0·25
4	8	3	1	1	0	0
5	13	5	4	2·5	1·5	2·25
6	10	5	2	2·5	−0·5	0·25
7	12	6	3	4	−1·0	1·00
8	20	9	6	5·5	0·5	0·25
					0	4·50

It will be noted that some of the values should be ranked equal, e.g. $y_1 = y_5 = 25$ should give ranks $Y_1 = Y_5 = $ 7th equal. In this case an average rank is used over the number of ranks that would have been used had they not been equal, e.g. $\dfrac{7+8}{2} = 7\cdot5$ or for x_2, x_7: $\dfrac{2+3}{2} = 2\cdot5$.

A check on the difference calculations is provided by the fact that $\Sigma\, d_i$ always equals zero.

$$\text{Now } \rho = 1 - \frac{6\,.\,4\cdot5}{8(64-1)} = 1 - 0\cdot0535 = 0\cdot947.$$

For a linear relationship, the calculated values of ρ and r will be similar and the significance of ρ may be tested in the same way as for r.

Note. The sign of ρ comes out automatically from the calculation. A proportional relation gives small $\Sigma\, d^2$ and therefore positive ρ. An inverse relation gives large Σd^2 and therefore negative ρ.

Thus on testing for correlation between turnover and unfulfilled orders, it is found that $\Sigma\, d^2 = 159\cdot5$ and that $\rho = -0\cdot85$. The reader may verify this for himself and also that $r = -0\cdot85$.

Example 10.5. Calculate the rank correlation coefficients for: (*a*) Example 10.2; (*b*) Example 10.3.

(*a*) Let rank of $x_i = R_i$

Let rank of $y_i = S_i$

TABLE 10.6

S_i	4	3	6	5	1	2	
R_i	5	3·5	6	3·5	1	2	*Total*
d_i	−1	−0·5	0	1·5	0	0	0·0
d_i^2	1	0·25	0	2·25	0	0	3·5

From Table 10.6:

$$\rho = 1 - \frac{6\,\Sigma\, d^2}{n(n^2-1)} = 1 - \frac{6\,.\,3\cdot5}{6\,.\,35}$$

$$= 1 - 0\cdot1 = 0\cdot9$$

(*b*)

TABLE 10.7

S_i	2·5	6·5	4·5	8	1	4·5	2·5	6·5	
R_i	1·5	3·5	5·5	7·5	1·5	3·5	5·5	7·5	*Total*
d_i	1	3	−1	0·5	−0·5	1	−3	−1	0
d_i^2	1	9	1	0·25	0·25	1	9	1	22·5

From Table 10.7:

$$\rho = 1 - \frac{6\,.\,22\cdot5}{8\,.\,63} = 1 - 0\cdot268$$

$$= 0\cdot73$$

10.6 CONFIDENCE LIMITS

Although the brand manager can estimate a value of turnover for a particular advertising appropriation from the regression equation, he knows that he is unlikely to make that turnover figure exactly. He may make more or less and would like to know the reliance that he can put on the estimate. In other words, how confident should he be in the estimate?

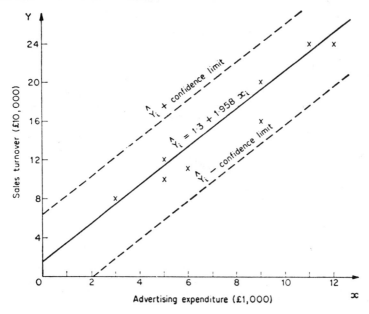

Figure 10.8   Approximate confidence limits

In applying the Least Squares Criterion, the assumption is made that the observations of y are evenly distributed about the line for a given x and that there should be an 'even' spread of points all the way up the line. The width of this spread governs the reliability of the estimate. In Chapter 2 it is stated that the standard deviation is a measure of spread. In the context of regression a comparable statistic known as the *Standard Error* of the estimate is calculated. It is a measure of the error remaining having fitted the line and is the standard deviation of the estimate (*see* Chapter 4).

$$\text{Standard Error, } S.E. = \sqrt{\left[\frac{n}{n-2}(1-r^2)\left(\frac{\Sigma y^2}{n} - \bar{y}^2\right)\right]}$$

It is convenient but not strictly accurate to look upon 95% confidence limits as the limits on either side of the estimate within which there is a 95% chance that the observation will fall. Thus confidence limits are based upon probability distributions and in order to use the convenient t-distribution it is necessary to make the further assumption that the y values are normally distributed about the line.

There are two possible formulae for confidence limits, one of which is an approximation to the other. The approximation formula makes the assumption that the band of confidence is parallel to the fitted line but this will only be

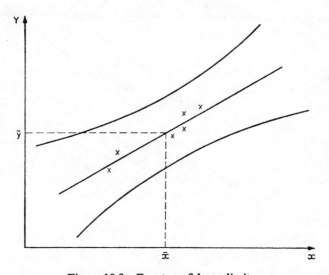

Figure 10.9   Exact confidence limits

justifiable if it is used within the range of the data as shown in Figure 10.8. The exact formula which is more complex gives a band of confidence which widens as the point at which one is estimating moves away from the central value $(\bar{x}, \bar{y})$. It is illustrated in Figure 10.9. To calculate 95% confidence limits both formulae require the value of t with $(n - 2)$ degrees of freedom which will be exceeded with a probability of 0·05, i.e. t_{n-2} 5%

(a) Approximation formula

95% Confidence limit $= \pm S.E. \times t_{n-2}$ 5%

For example if the brand manager above wishes to calculate the approximate 95% confidence limit for use within the range of the advertising expenditure data, he first calculates the standard error of the estimate:

$$S.E. = \sqrt{\left[\frac{8}{6}(1 - 0.953^2)\left(\frac{2,352}{8} - 16^2\right)\right]}$$
$$= 2.16$$

He then looks up the value of t_6 $5\% = 2\cdot447$.
The confidence limit is then $\pm 2\cdot447 \times 2\cdot16 = 5\cdot286$.
The estimating equation now becomes:

$$Y_i = 1\cdot3 + 1\cdot958\ x_i \pm 5\cdot286,\ \text{or}$$
Turnover $= 13,000 + 19\cdot58$. Advertising $\pm 52,860$.

For an advertising appropriation of £4,000, his estimate is £91,320 $\pm$ £52,860, i.e. there is a 95% chance that turnover will be between £38,460 and £144,180. The confidence limits are shown in Figure 10.8. The limits are fairly wide in this example due to the high probability level set (i.e. 95%) and the width of scatter of points about the line. The level of confidence required is set by the user.

(b) *Exact formula*
    95% Confidence limit =

$$\pm \sqrt{\left[(S.E.)^2 \left(1 + \frac{1}{n} + \frac{(x_0 - \bar{x})^2\ n}{n\ \Sigma\ x^2 - (\Sigma\ x)^2} \right) \right]} \times t_{n-2}\ 5\%$$

where x_0 is the value from which an estimate of y is to be obtained.
For example, if the brand manager requires an estimate of expected turnover for an advertising appropriation of £15,000, his confidence limits are given by:

$$\pm \sqrt{\left[2\cdot16^2 \left(1 + \frac{1}{8} + \frac{(15 - 7\cdot5)^2\ .\ 8}{8\ .\ 522 - 60^2} \right) \right]} \times 2\cdot447 \times 10,000$$
$$= \pm 2\cdot98 \times 2\cdot447 \times 10,000$$
$$= \pm £72,975$$

Example 10.6. Calculate the confidence limits for the equations of (a) Example 10.2 estimating turnover from making 5 calls; (b) Example 10.3 estimating % increase in sales if commission rates are increased by 80%.

(a) Standard Error, $S.E. = \sqrt{\left[\frac{6}{4}(1 - 0\cdot884^2) \left(\frac{706}{6} - \frac{64^2}{36} \right) \right]}$
$$= 1\cdot13$$
$$t_4\ 5\% = 2\cdot776$$

Approximate Confidence limit for $Y = \pm 1\cdot13 \times 2\cdot776$
$$= \pm 3\cdot14$$
$$\text{but } Y = \frac{\text{Sales turnover}}{100}$$

and thus an estimate for sales turnover has a 95% confidence limit of $\pm £314$.

    i.e. Sales turnover $= 729\cdot2 + 67\cdot5$ (No. of calls) ± 314

For 5 calls, there is a 95% chance that turnover will be between £752·7 and £1,380·7.

(b) $S.E. = \sqrt{\left[\frac{8}{6}(1 - 0.73^2)\left(\frac{404}{8} - 49\right)\right]} = 0.97$

$t_6\ 5\% = 2.447$

$x_0 = 8$

Exact confidence limit $= \pm^1\sqrt{\left[0.97^2\left(1 + \frac{1}{8} + \frac{(8 - 4.5)^2\ 8}{8\ .\ 172 - 36^2}\right)\right]} \times 2.447$

$\qquad\qquad = \pm 3.64\%$

i.e. % increase in sales $= 3.4 + 0.08 \times 8 \pm 3.64$

$\qquad\qquad\qquad\qquad = 4.04 \pm 3.64$

There is a 95% chance that sales will increase by between 0.40% and 7.68% if commission rates are increased by 80%.

These confidence limits are fairly wide due to the marginal significance of the correlation coefficients, the small quantity of data and the probability level set (95%).

10.7 NON-LINEAR REGRESSION

As mentioned the pattern indicated by a scatter diagram is not always linear and it may be necessary to choose some other equation to fit to the data. This equation may be of exponential form or quadratic such as ellipsoidal or parabolic. It is still possible to fit the equation by the method of Least Squares although in some cases one may transform the non-linear function into a linear function before applying the Least Squares method, e.g. by taking logarithms in the case of an exponential function:

$$y = a\,e^{bx}$$

gives

$$\log_e y = \log_e a + bx$$

which is linear in $\log_e y$ and x. However the transformation in this case only gives an approximation to the true parameters since it gives a disproportionate weight to observations of low values of y. In other cases a convenient transform cannot be found and it is necessary to use the Least Squares method on the basic function which can lead to complicated equations for the constants.

The discussion so far has established a relationship between a dependent variable and only one independent variable. This may be an unsatisfactory model to explain the situation and one might believe that y depends on two or more independent variables, e.g. that turnover depends on both advertising and market share. Here a multivariate model is fitted such as:

$$y = a + bx_1 + cx_2$$

Where: $y =$ turnover

$\quad\ x_1 =$ advertising expenditure

$\quad\ x_2 =$ market share

The equation is fitted by the Least Squares method which minimizes $\Sigma\,(y_i - a - bx_{1_i} - cx_{2_i})^2$ giving equations for a, b, and c.

To fit more than two independent variables, or even a more complex equation in just two variables, it is often necessary to use a multiple regression computer package to carry out the calculations as indicated in Chapter 12.

10.8 SUMMARY

If a scatter diagram indicates that a linear relationship might exist between two variables, y and x, the correlation between them should first be calculated and tested for significance before calculating the equation of the line.

$$\text{Correlation coefficient, } r = \frac{n\,\Sigma\,xy - \Sigma\,x\,.\,\Sigma\,y}{\sqrt{[(n\,\Sigma\,x^2 - (\Sigma\,x)^2)(n\,\Sigma\,y^2 - (\Sigma\,y)^2)]}}$$

or as an approximation the rank correlation coefficient $\rho = 1 - \dfrac{6\,\Sigma\,d^2}{n(n^2 - 1)}$

where $d_i = R_i - S_i$ and R and S are the ranks of x and y.

Test of significance:

Calculate $t = r\sqrt{\dfrac{n-2}{1-r^2}}$ and compare with the value of t with $(n-2)$ degrees of freedom at the required significance level from Table T.2. If t is greater than that tabulated, r (or ρ) is significant and there is some justification for calculating the equation of the line.

Regression of y on x:

$$Y_i = a + bx_i$$

$$b = \frac{n\,\Sigma\,xy - \Sigma\,x\,.\,\Sigma\,y}{n\,\Sigma\,x^2 - (\Sigma\,x)^2}$$

$$a = \bar{y} - b\bar{x}$$

Standard error of Y_i, $S.E. = \sqrt{\left[\dfrac{n}{n-2}(1 - r^2)\left(\dfrac{\Sigma\,y^2}{n} - \bar{y}^2\right)\right]}$

Approximate 95% Confidence limits $= \pm S.E.\ t_{n-2}5\%$

Exact 95% confidence limits $= \pm\sqrt{\left[(S.E.)^2\left(1 + \dfrac{1}{n} + \dfrac{(x_0 - \bar{x})^2\,.\,n}{n\,\Sigma\,x^2 - (\Sigma\,x)^2}\right)\right]}$

$\times\ t_{n-2}\,5\%$

EXERCISES

1. The average weekly weight of a group of steers were recorded in a fattening trial. Calculate the linear regression relationship between the number of weeks on experiment and the average weight of a steer.

Number of weeks on trial	Average weight of steers (kg)
1	195
2	203
3	210
4	217
5	228
6	238
7	244
8	253
9	263
10	265

(*From IOS, Part 1*)

2. Define (*a*) (Product-moment) correlation coefficient
       (*b*) Rank correlation coefficient
Illustrate your answer with reference to the following data:

x	1	2	3	4	5
y	1·5	4·6	2·8	6·2	4·9

(*From IOS, Part 1*)

3.

Deliveries of cookers
Monthly averages
Total in thousands

Year	Gas	Electricity
1959	66·8	39·5
1960	54·2	42·5
1961	46·5	40·9
1962	56·5	40·9
1963	54·4	40·6
1964	65·0	47·5
1965	60·9	44·5
1966	64·2	45·5
1967	63·3	51·4

Calculate the coefficient of correlation between the two series and comment on the result.

(*IM, Part 2*)

4. The following table gives the order in which two judges X and Y placed eight contestants in an essay competition:

	A	B	C	D	E	F	G	H
X	1	5	3	4	2	7	6	8
Y	3	7	6	1	8	4	2	5

Calculate the coefficient of rank correlation and state any conclusions you draw from it.

(From IOS, Part 2)

5. (*a*) Explain the logic of fitting a straight line to data by means of the Least Squares Criterion.

(*b*) How may the significance of a rank correlation coefficient be tested?

(*c*) The following are paired data on the price of a share of company X and the Financial Times Ordinary Share Index quoted on days picked at random.

Share price (p)	75	47	84	72	66	70	59	73	75	87
F.T. Index	319	316	388	335	382	339	344	355	358	399

Is there a significant correlation between the share price and the F.T. Index?

(University of Strathclyde)

6. The following quarterly data on cinema admissions and TV licences issued has been gathered over a three year period.

Quarter	Cinema admissions (weekly average in millions)	TV licences issued (per 1,000 population)
1	11·0	12
2	10·0	18
3	10·0	24
4	9·4	37
5	10·5	52
6	9·7	64
7	9·4	69
8	9·3	81
9	9·9	98
10	9·3	101
11	9·0	106
12	8·6	119

What conclusions can be drawn from this data?

(University of Strathclyde)

7. A set of experiments was carried out on a chemical plant to test for a relationship between the percentage yield of the product and the reaction temperature in degrees Centigrade. The following data was obtained from eight experiments:

Temperature in °C	30	40	50	60	30	40	50	60
Yields as %	60	80	70	90	50	70	60	80

Calculate the correlation coefficient for this data. Does it indicate a significant relationship between yields and temperature at the 5% level of significance? Calculate the percentage yield to be expected using a reaction temperature of 55°C giving the 95% confidence limits.

Is there justification for saying that increasing the temperature *causes* an increase in the yield?

(*University of Strathclyde*)

11. *Demand Forecasting*

11.1 INTRODUCTION

Decisions are made in the light of the information available which often has to take the form of estimates of future values of variables such as demand for the products, commodity prices, incomes growth, etc. In particular estimates of future demands are necessary for planning production, stock control policies, cash budgeting, and capital expenditure.

The reader should note that demand is a different variable to sales. Sales are those demands which are satisfied and the recorded sales figures may well be less than the total demand for the product. Strictly, predictions of future demands are required for planning purposes, and the company's activities govern the extent to which those demands are satisfied and hence the level of future sales. Thus the implication is that forecasts of demand should be based on past demand data (and any other relevant factors) and *not* on past sales figures. Unfortunately many companies do not keep records of unfulfilled orders and the forecasting system must be started off by using past sales data alone. However, once this situation is recognized steps should be taken to collect demand data in future. The following discussion is confined to demand forecasting but the methods are applicable to forecasting many other variables.

Estimates of future demand may be made in one of two ways or as a combination of both.

Subjective estimates. Subjective estimates are based upon the considered judgment of managers which in turn is based upon their intuition and knowledge of the market environment. The knowledge is usually built up by experience and good estimates are often produced. Apart from adding to his knowledge little outside assistance can be given to the manager making subjective estimates. This form of estimating is most commonly used on products for which there is a volatile market requiring considerable promotional activity. These products are often the high volume/high contribution lines in the company but there may be relatively few of them. Long-term predictions are often obtained as subjective estimates because of the lack of numerical information about the future.

Objective estimates. Estimates are made by mathematically analysing past data (or facts) on demand and related variables and then extrapolating the results of the analysis into the future. The analysis is often termed *time series analysis.* Mathematical forecasting is most effective for products in a stable market since there are obvious dangers in extrapolating past patterns of unstable markets into the future. Fortunately a considerable number of products fall into the stable market category and thus lend themselves to computerized forecasting systems. Objective estimating procedures are most used for producing short-term forecasts.

In practice objective forecasts are often subjectively assessed for validity before being used in the decision-making process. This chapter is concerned only with objective forecasts and these fall into one of two categories.

11.1.1 EXTRINSIC FORECASTS

Extrinsic forecasts are based on past data of both demand and other related variables. For example demands for motor-car spare parts may depend on prior sales of cars and the age of the particular model. A forecasting equation for demand might be formulated in terms of past spares demand patterns, car sales, and age of the model. The essence of the analysis is to find the mathematical relationship between each variable's time series and the methods of regression or multiple regression (*see* Chapters 10 and 12) may be appropriate.

Extrinsic forecasting models tend to be somewhat cumbersome and awkward to update rapidly. It is usually simpler to use intrinsic models provided that the forecasts so produced are adequately accurate.

11.1.2 INTRINSIC FORECASTS

Intrinsic forecasts are based purely on past data of demands. The analysis consists of identifying the pattern of the data as shown in a time series. The

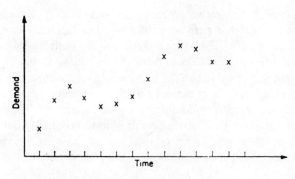

Figure 11.1   Time series of demand

example in Figure 11.1 indicates that demand follows a rising trend with marked *seasonal* fluctuations. In practice time series rarely show such a clear pattern and there are usually random fluctuations imposed on the basic pattern.

The methods of *moving averages* and *exponential smoothing* are available for forecasting trend models and they are first illustrated in the context of a constant demand model. The methods are primarily for use in short-term forecasting.

11.2 CONSTANT MODEL

A constant demand model is a special case of a trend model whose slope is zero. Figure 11.2 illustrates such a model and the points are to be taken as past data on actual demand. Each plot represents the total demand since the last plot and is

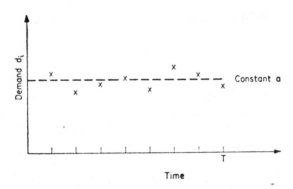

Figure 11.2   Constant demand model

plotted at the end of its time period for convenience. It will be noted that the plots fluctuate at random about some constant level, *a*. This situation could occur in practice where there was no significant growth or decline in demand for the product in the time periods being considered, i.e. a very stable market.

The graph is represented mathematically by:

$$d_i = a + \varepsilon_i$$

where d_i = actual or observed demand at time i (i.e. the total demand in the period ending at time i)

a = basic constant level of demand

ε_i = 'noise' term or variation of demand from a at time i. The noise term, ε_i, is a random variable with zero mean and unknown variance.

In using this demand model it is assumed that the next observation will have a value near the level a. That observation may be split into two parts, the constant a

and the noise term ε_t. Since ε_t is a random variable, its next value cannot be forecast and therefore the only forecast of demand that can be made is an estimate of a. When that observation does eventually occur it is unlikely to be at precisely the level estimated for a. The deviation may be due just to chance or due to a slight change in the basic demand level. The estimate of the basic demand level is therefore updated, i.e. the estimate of a is updated and a new forecast produced. The forecasting methods discussed below can in fact be used when the basic demand level is slowly changing, i.e. a very slow trend. They are not suitable in situations where the demand level changes significantly each period.

In this chapter the forecast made at time T(now) for some other point in time t is denoted by $y_{T,t}$. Thus $y_{T,t}$ is the estimate of a. For a constant model the point in time t for when the forecast is made does not have to be specified since the forecast made now applies to the next period or any other period in the future until such time as it is updated. For any other model time t must be made explicit (*see* Section 11.3).

11.2.1 MOVING AVERAGES

An estimate of a may be obtained by taking it as the average level of past demands. It is possible to calculate the average of all the past demand observations but in practice the average is calculated from the last N observations so that it is based on recent history and shows up changes in the basic demand level. For example if the last four observations were 125, 113, 108, and 120, then their average is 116·5 and this is taken as the forecast for the fifth period. Suppose that the fifth observation now occurs and is 125. Then the average of the last four observations is still 116·5 and this is taken as the forecast for the sixth period. Here the average has been calculated arbitrarily over only four periods. The process can be put mathematically as:

$$y_{T,t} = \frac{1}{N}(d_T + d_{T-1} + d_{T-2} + \ldots + d_{T-N+1})$$
$$= \frac{1}{N} \sum_{i=T-N+1}^{T} d_i \qquad [11.1]$$

It should be noted that because the time t for when the forecast is made does not have to be specified for a constant model, t does not appear on the right-hand side of the equation.

d_T is the latest demand, i.e. at now, d_{T-1} is the previous demand and so on. Assuming that d_5 has not yet occurred in Figure 11.3, the forecast made at time 4 using a four-period average is:

$$y_{4,t} = \tfrac{1}{4}(d_4 + d_3 + d_2 + d_1)$$

At time 5, a new observation, d_5, occurs and the updated forecast is:

$$y_{5,t} = \tfrac{1}{4}(d_5 + d_4 + d_3 + d_2)$$

In order to show that the lower limit of the summation sign in Equation 11.1 is correct consider this case of T taken to be 5 and N, the period of the average, to be 4, then:

$$T - N + 1 = 5 - 4 + 1 = 2$$

The value of 2 corresponds to d_2 in the equation for $y_{5,t}$. Thus the lower limit of the summation sign is correct. At each time period the updating calculation adds

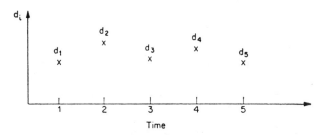

Figure 11.3 Moving averages

in the latest demand and drops the oldest. The average thus moves on and is known as an N period moving average.

An alternative form of the moving average equation which illustrates the moving effect is:

$$y_{T,t} = y_{T-1,t} + \frac{1}{N}(d_T - d_{T-N}) \qquad [11.2]$$

where $y_{T-1,t}$ is forecast made in the previous period.

The equivalence of Equations 11.1 and 11.2 is shown in Appendix F. The convenience of this formula is illustrated by the following example where the time periods are quarterly:

$$i = \quad 1 \quad\quad 2 \quad\quad 3 \quad\quad 4$$
$$d_t = \quad 125 \quad 113 \quad 108 \quad 120$$

$$y_{4,t} = \tfrac{1}{4}(125 + 113 + 108 + 120) = 116\cdot5$$
$\qquad$ = forecast made in 4th quarter and plotted at time 4 in Figure 11.4 and which applies to time 5, 6, 7, 8, etc.

Now $d_5 = 125$

then $y_{5,t} = y_{4,t} + \tfrac{1}{4}(d_5 - d_1)$
$\qquad\quad = 116\cdot5 + \tfrac{1}{4}(125 - 125) = 116\cdot5$

The forecasts for five quarters are shown in Figure 11.4 where the data is:

$i =$	1	2	3	4	5	6	7	8
$d_i =$	125	113	108	120	125	112	110	115
$y_{T,t} =$				116·5	116·5	116·25	116·75	115·5

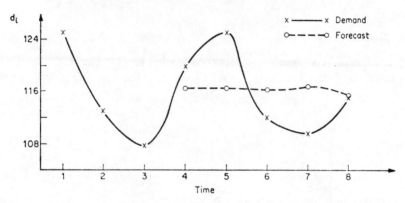

Figure 11.4   Moving averages example

Moving averages have two main disadvantages:

(*a*) They require N pieces of data in order to make a forecast. This could be time-consuming and expensive in terms of data handling and storage.

(*b*) They give equal weight to the last N observations and zero weight to any older observations. This is an arbitrary cut-off point which depends on the size of N.

The size of N also affects the speed with which the forecasts will adjust to any change in the basic demand level. Small N gives all the weight to the very latest observations and the forecasts therefore follow the path of demand closely but lagging behind it and they may be following freak observations. Large N takes many observations into account and thereby smoothes out the effect of freak observations but, by the same token, also reduces the sensitivity of the forecast to a change in the demand pattern. The choice of N must therefore be a subjective compromise governed by the likely level of stability of the demand pattern. Its value is often chosen by experience.

The first stage of the seasonal forecasting method discussed in Section 11.4 requires the data to be deseasonalized. If the observations are monthly and $N = 12$, moving averages have the advantage of smoothing out the seasonal effects. Similarly a 4 period moving average deseasonalizes quarterly data. Frequently the wish to deseasonalize data for seasonal forecasting or for some index numbers governs the size of N.

11.2.2 EXPONENTIAL SMOOTHING

Since it is reasonable to assume that the oldest observations have least effect on or connection with the future, a more realistic method is to give a steadily decreasing weight to older observations. In this way the forecast should adapt more rapidly to changes in demand. It is also more reasonable to take the average over all observations in time and not cut off observations older than N periods. A *weighting factor* β is used which is a fraction lying between zero and one. An observation r periods old has a weight proportional to β^r, i.e.:

Observation at time T is 0 periods old and has a weight $\beta^0(1 - \beta) = (1 - \beta)$

   ,,   ,, $(T - 1)$,, 1   ,,   ,,   ,, $\beta^1(1 - \beta) = \beta(1 - \beta)$

   ,,   ,, $(T - 2)$,, 2   ,,   ,,   ,, $\beta^2(1 - \beta)$

   ,,   ,, $(T - r)$,, r ,,   ,,   ,, $\beta^r(1 - \beta)$

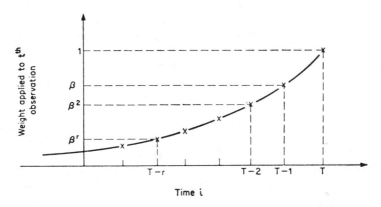

Figure 11.5   Exponential weighting

The weights are shown graphically in Figure 11.5. The relative weights 1, β, β^2, ..., β^r, ... lie on an exponential curve and are said to decrease exponentially.

A weighted average may now be calculated whose equation is:

$$y_{T,t} = \beta \cdot y_{T-1,t} + (1 - \beta) \cdot d_T \tag{11.3}$$

i.e. Forecast $= \beta \times$ previous forecast $+ (1 - \beta) \times$ latest demand.

Thus the forecast made now is calculated in terms of the previous forecast and the latest demand. The previous forecast in fact contains or is made up of all the past information up to time $(T - 1)$. As before the average is moving forward and taking in the latest observation. It is known as an Exponentially Weighted Moving Average and the process is known as *Exponential Smoothing*. It has the advantage of only requiring to store two pieces of data to make a forecast.

Equation 11.3 is similar to Equation 11.2. The exponential smoothing equivalent of Equation 11.1 is derived in Appendix F and it shows that each observation has a weight which depends on its age.

A high value of β gives most weight to the previous forecast and thus smoothes out excessive noise fluctuations but gives a slow response to a change in the demand pattern. A low value of β gives most weight to the latest observation and the forecast will therefore follow actual demands more closely but a few periods behind. The forecast may then be following chance fluctuations. Thus it is necessary to use a compromise value for β which takes account of the stability of the basic demand pattern. A value of between 0·8 and 0·9 is usually chosen for β on the assumption that the constant model is the correct model for the given situation, i.e. demand is stable.

Equation 11.3 indicates that a starting value is required for the previous forecast when the system is set up. However the system is not too sensitive to relatively small errors in that value. A sufficiently accurate estimate may be obtained by taking an average value over, say, the first two-thirds of the available past data and then running the system in over the remaining one-third.

Using the example data in Section 11.2.1, suppose 'now' is at time $i = 6$ and thus the data available to date is:

$i =$	1	2	3	4	5	6(Now)
$d_i =$	125	113	108	120	125	112

The first two-thirds of the data $(i = 1$ to $4)$ is used to obtain a starting value for the previous forecast, i.e.:

$$y_{4,t} = \tfrac{1}{4}(125 + 113 + 108 + 120) = 116·5$$

The system is now run in over the remaining two periods using, say, β = 0·8. Then:

$$y_{5,t} = \beta \cdot y_{4,t} + (1 - \beta) \cdot d_5$$
$$= 0·8 \times 116·5 + 0·2 \times 125$$
$$= 118·2$$
$$y_{6,t} = \beta \cdot y_{5,t} + (1 - \beta) \cdot d_6$$
$$= 0·8 \times 118·2 + 0·2 \times 112$$
$$= 117·0$$

$y_{6,t}$ is the forecast made now for all time in the future until a new demand is observed and the forecast can be updated, i.e. when $d_7 = 110$. Then:

$$y_{7,t} = 0\cdot8 \times 117\cdot0 + 0\cdot2 \times 110$$
$$= 115\cdot6$$

When $d_8 = 115$, then $y_{8,t} = 115\cdot5$

These values are shown in Figure 11.6 where the forecasts are plotted against the time at which they are made since they can apply to any time in the future. Note that the forecasts are different to those obtained by moving averages and in this particular example are marginally worse. However that is not a generalization!

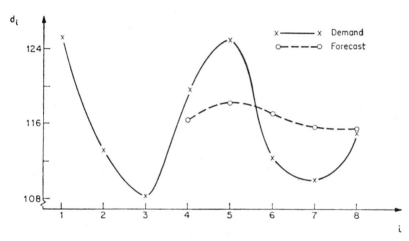

Figure 11.6   Exponential smoothing

It should be noted that the assumption that older observations have less relevance to the future than recent observations is not valid in situations where demands are seasonal. Exponential smoothing is not suitable for deseasonalizing data in the same way as moving averages since the method gives unequal weights to different parts of the cycle.

In literature on forecasting one often comes across Equation 11.3 in a different form containing α which is known as the *smoothing constant* and equals $(1 - \beta)$, i.e.:

$$\alpha = 1 - \beta$$

The equation becomes:

$$y_{T,t} = y_{T-1,t} + \alpha \cdot (d_T - y_{T-1,t})$$

which has an intuitive argument behind it. Consider Figure 11.7. A forecast $y_{T-1,t}$ was made at time $(T - 1)$. The observation at time T is d_T. What should the forecast at time T be? There are two extremes to the argument.

(*a*) The observation d_T was a freak and therefore the new forecast should equal the old, i.e.:

$$y_{T,t} = y_{T-1,t}$$

(*b*) The old forecast was completely wrong and the latest observation is truly representative of the basic situation and therefore the new forecast should equal the latest observation, i.e.:

$$y_{T,t} = d_T$$

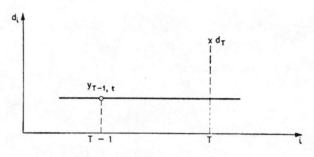

Figure 11.7

A compromise would be to make the new forecast lie somewhere between the old forecast and the latest observation. The difference between these two is given by:

$$d_T - y_{T-1,t}$$

and is known as the one-step ahead error, e_T. The new forecast should therefore equal the old forecast plus some fraction, α, of the one-step ahead error, i.e.

$$\begin{aligned} y_{T,t} &= y_{T-1,t} + \alpha \cdot e_T \\ &= y_{T-1,t} + \alpha \cdot (d_T - y_{T-1,t}) \end{aligned}$$

11.3 LINEAR TREND MODEL

Demand is following a reasonably steady growth or decline pattern, i.e. a linear trend. The basic model is given by:

$$d_i = a + b \times i + \varepsilon_i$$

where b = slope of the trend and time i is measured from some arbitrary origin such as January 1973.

To make a forecast it is now necessary to estimate values for a and b, i.e. the estimates of a and b are updated after each new observation. It is also necessary to specify the time for which the forecast is being made. The forecast $y_{T,t}$ is given by:

$$y_{T,t} = a_T + b_T \cdot (t - T) \qquad\qquad [11.4]$$

where a_T = value of a calculated at time T
b_T = value of b calculated at time T.

The trend situation is shown in Figure 11.8.

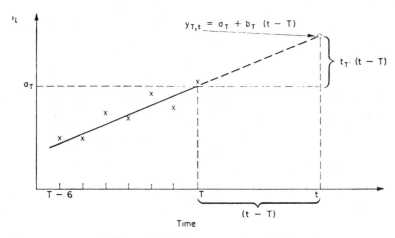

Figure 11.8   Linear trend model

11.3.1 MOVING AVERAGES

It will be noted that the above forecasting Equation 11.4 is of the form: $y = a + b \cdot x$ where the variable $(t - T)$ has replaced x. Therefore a straight line could be fitted to the data by the regression method using the last N observations. The regression method is an averaging process and if only the last N observations are used each period, it becomes a moving average method. As such it still has the disadvantages of the constant model moving average method and is also cumbersome to update. Therefore only an exponential smoothing method is considered.

11.3.2 EXPONENTIAL SMOOTHING

As before, older observations are given progressively less weight and a line may be fitted by the method of weighted or discounted least squares. (The proof is complex and therefore omitted but Brown's method of double smoothing, Reference 3 gives exactly equivalent results.) This gives values to the parameters a_T and b_T in terms of their previous values and the latest one step ahead error:

$$a_T = a_{T-1} + b_{T-1} + (1 - \beta^2) . e_T$$
$$b_T = \qquad\qquad b_{T-1} + (1 - \beta)^2 . e_T$$

where $e_T = d_T - y_{T-1,T}$ [11.5]

These values are then put back in the forecasting equation. The method requires starting values for a_{T-1} and b_{T-1} which may be obtained by calculating a regression line (or fitting a line by eye) to the first two-thirds of available data; b_{T-1} is the slope of that line and a_{T-1} is the value estimated from the line at time $(T - 1)$. As before these initial values are modified during the 'running in' period over the remaining one-third of the data. β is normally taken to have a value of about 0·9.

Example. A new product was marketed nine months ago and data on monthly turnover in £1,000 units is available for that period. A forecasting system is required giving forecasts for next month and for three months' time.

$$i = 1 \quad 2 \quad 3 \quad 4 \quad 5 \quad 6 \quad 7 \quad 8 \quad 9(\text{now})$$
$$d_i = 3 \quad 3 \quad 3 \quad 6 \quad 5 \quad 8 \quad 7 \quad 9 \quad 11$$

The first six months are used to calculate starting values for a_{T-1} and b_{T-1}. The regression equation calculated from those values is:

$$d_i = 1·27 + 0·97 \times i$$

Thus b_6 may be taken as 0·97 or approximately 1·0 and a_6 is given by $1·27 + 0·97 \times 6 = 7·09$ or roughly 7·0. These are the starting values and the system is run in over months 7 to 9. For ease of calculation β is taken to be $\frac{1}{2}$.
Thus $a_6 = 7$, $b_6 = 1$, $\beta = \frac{1}{2}$.

$$y_{6,7} = a_6 + b_6 \times 1 = 7 + 1 = 8$$

and $(1 - \beta^2) = 1 - 0·25 = 0·75$
$\qquad (1 - \beta)^2 = 0·5^2 \qquad = 0·25$

At time 7 the error is $d_7 - y_{6,7} = 7 - 8 = -1 = e_7$

then $a_7 = a_6 + b_6 + (1 - \beta^2) . e_7$
$\qquad\quad = 7 + 1 + 0·75 \times (-1)$
$\qquad\quad = 7·25$
$\qquad b_7 = b_6 + (1 - \beta)^2 . e_7$
$\qquad\quad = 1 + 0·25 \times (-1)$
$\qquad\quad = 0·75$
and $y_{7,8} = a_7 + b_7 = 7·25 + 0·75 = 8·0$

At time 8 the error is $e_8 = 9 - 8 = 1$

$$a_8 = 7.25 + 0.75 + 0.75 \times 1 = 8.75$$
$$b_8 = \qquad\quad 0.75 + 0.25 \times 1 = 1.0$$

and $y_{8,9} = 8.75 + 1.00 = 9.75$

At time 9, $e_9 = 11 - 9.75 = 1.25$

$$a_9 = 9.75 + 0.75 \times 1.25 = 10.69$$
$$b_9 = 1.00 + 0.25 \times 1.25 = 1.31$$

and $y_{9,10} = 10.69 + 1.31 = 12.00$

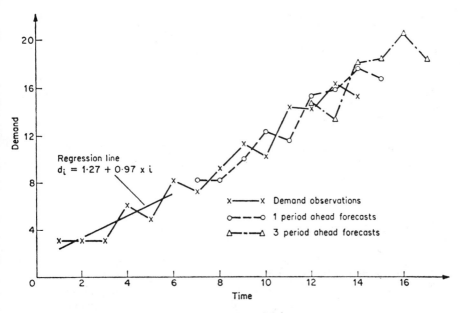

Figure 11.9 Trend model forecasts

A forecast can be made now (i.e. $T = 9$) for any specified time in the future, say 3 months ahead, i.e. $t = 12$. Thus:

$$y_{9,12} = a_9 + b_9 \cdot (12 - 9)$$
$$= 10.69 + 1.31 \times 3$$
$$= 14.62$$

At each succeeding month, another turnover figure is recorded and the forecasts updated. The whole procedure may be simplified by use of a work sheet such as Table 11.1. The observations and forecasts (plotted at the time for when they are made) are shown in Figure 11.9.

TABLE 11.1

Work sheet for trend model forecasts

T	d_T	e_T	a_T	b_T	$y_{T,T+1}$	$y_{T,T+3}$
6	8		7·0	1·0	8·0	
7	7	−1·0	7·25	0·75	8·0	
8	9	1·0	8·75	1·0	9·75	
9	11	1·25	10·69	1·31	12·00	14·62
New data						
10	10	−2·00	10·5	0·81	11·31	12·93
11	14	2·69	13·33	1·48	14·81	17·77
12	14	−0·81	14·21	1·28	15·49	18·05
13	16	0·51	15·87	1·41	17·28	20·10
14	15	−2·28	15·57	0·84	16·41	18·09

11.4 SEASONAL MODEL

Finally the situation is considered in which demand is believed to follow a seasonal pattern usually superimposed on a trend. A trigonometric model for such a pattern is given by:

$$d_i = a + b . i + c \sin w i + g \cos w i + \varepsilon_i$$

where w = angular frequency.

The parameters a, b, c, and g have to be estimated and these estimates may be obtained by applying the discounted least squares method to the above model. However the calculations become very complex and unless very accurate forecasts are required (usually for high value products) the method may be discarded in favour of the following simpler approach which makes use of a trend model with an additive seasonal factor.

The essentials of the method are first to deseasonalize the data thereby identifying the underlying trend. The differences between the observations and this trend are taken as measures of the seasonal effects. A forecast is made by extrapolating the trend line (using the method of Section 11.3.2) and superimposing the appropriate seasonal deviation. The method is explained by means of an example.

Assume that a company has three years quarterly data on demand in units of 1,000 packages.

$i =$	1	2	3	4	5	6	7	8	9	10	11	12
$d_i =$	26	15	11	24	30	18	17	23	31	20	19	27

It uses the first two years (quarters 1 to 8) to calculate the seasonal effects and a basic trend. The system is to be 'run in' over the third year.

The data is first *deseasonalized* by calculating a 4 period moving average. Strictly these averages fall between the second and third observations of each set of four. They are plotted as such in Figure 11.10. A trend line is fitted to these points either by eye or by regression depending on the accuracy required. The

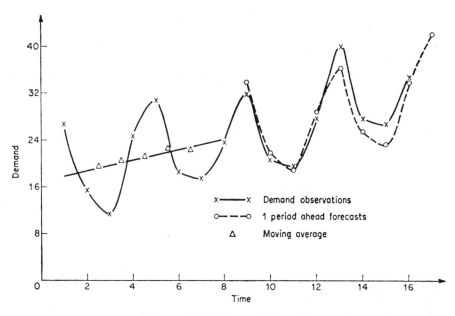

Figure 11.10 Seasonal forecasts

deseasonalized demands (x_i) may now be read off (or calculated from) the trend line against each quarter and the figures are:

$i =$	1	2	3	4	5	6	7	8
M.A. $=$		19	20	20·75	22·25	22		
$x_i =$	17·9	18·6	19·4	20·2	21·1	22·0	22·8	23·6

The deviations of each quarter's demand from the trend line or deseasonalized demands are tabulated (Table 11.2). The average deviation per quarter type is then calculated and the sum of these average deviations should be zero otherwise a bias is introduced. If they do not sum to zero, each must be adjusted by subtracting the average 'average deviation'. In this example their sum is −0·8 and therefore 0·2 is added to each individual deviation. The column of corrected deviations is then a measure of the effect of seasonality.

TABLE 11.2
Additive seasonal deviations

Quarter	Year		Average	Corrected
j	1	2	deviation	deviation, p_j
1	8·1	8·9	8·5	8·7
2	−3·6	−4·0	−3·8	−3·6
3	−8·4	−5.8	−7·1	−6·9
4	3·8	−0·6	1·6	1·8
			−0·8	0·0

Let the corrected deviation for the jth quarter be p_j. Let the forecast made at time T for time t ahead which is of the jth type of quarter be $y_{T, t(j)}$. Then:

$$y_{T, t(j)} = a_T + b_T . (t - T) + p_j \qquad [11.6]$$

The updating equations for a_T and b_T are the same as for the linear model (i.e. Equation 11.5) but now the one step ahead error is given by:

$$e_T = d_T - y_{T-1, T(j)}$$
$$= d_T - (a_{T-1} + b_{T-1} + p_j)$$

Initial values for the trend line may be obtained from the deseasonalized demands, x_t, i.e.:

$$a_8 = x_8 = 23·6$$
$$b_8 = \frac{23·6 - 17·9}{7} = 0·82$$

Then the first forecast made at time 8 for time 9 which is a quarter of type 1, i.e. the first quarter of the third year, is:

$$y_{8,9(1)} = a_8 + b_8 + p_1$$
$$= 23·6 + 0·82 + 8·7 = 33·12$$

The next observation at time 9 is $d_9 = 31$ and the parameters a_T and b_T are updated using a value of β in this example of 0·9, i.e.:

$$(1 - \beta^2) = (1 - 0·9^2) = 0·19 \text{ or approximately } 0·2$$
$$(1 - \beta)^2 = (1 - 0·9)^2 = 0·01$$

Then $e_9 = d_9 - y_{8,9(1)}$
$$= 31 - 33·12 = -2·12$$
$$a_9 = a_8 + b_8 + (1 - \beta^2) . e_9$$
$$= 23·6 + 0·82 + 0·2 \times (-2·12)$$
$$= 24·0$$
$$b_9 = b_8 + (1 - \beta)^2 . e_9$$
$$= 0·82 + 0·01 \times (-2·12)$$
$$= 0·80$$

and

$$y_{9,\,10(2)} = a_9 + b_9 + p_2$$
$$= 24{\cdot}0 + 0{\cdot}8 - 3{\cdot}6$$
$$= 21{\cdot}2$$

The calculations may be laid out on a work sheet as in Table 11.3. The system is run in up to time 12 at which point a forecast is made for time 13, i.e. 35·54 is forecast. Thereafter new demand data arrives at quarterly intervals and the forecast is updated. The second half of Table 11.3 illustrates the continuing process and all the data and forecasts have been plotted in Figure 11.10.

TABLE 11.3

Work sheet for seasonal forecasting

Period T	d_T	e_T	a_T	b_T	$a_T + b_T$	p_{T+1}	$y_{T,\,T+1(J)}$
9	31	−2·12	24·00	0·80	24·80	−3·6	21·20
10	20	−1·20	24·56	0·79	25·35	−6·9	18·45
11	19	0·55	25·46	0·80	26·26	1·8	28·06
12	27	−1·06	26·05	0·79	26·84	8·7	35·54
13	39	3·46	27·53	0·82	28·35	−3·6	24·75
14	27	2·25	28·80	0·84	29·64	−6·9	22·74
15	26	3·26	30·29	0·87	31·16	1·8	32·96
16	34	1·04	31·37	0·88	32·25	8·7	40·95

It is possible to make forecasts for any period ahead. For example the three month ahead forecast at time 12 is:

$$y_{12,\,15(3)} = a_{12} + b_{12} \times (15 - 12) + p_3$$
$$= 26{\cdot}05 + 0{\cdot}79 \times 3 - 6{\cdot}9$$
$$= 21{\cdot}52$$

Occasionally, maybe annually, the seasonal deviations, p_j, should be updated. The method can also be applied to monthly data in which case a 12 month moving average is used for deseasonalizing the data.

An alternative simple method of seasonal forecasting is that using seasonal ratios. These ratios are calculated for each period type as the ratios of the observation to the deseasonalized demand, i.e. $q_j = d_j/x_j$ for the jth period type.

The ratios may be calculated over two or three years of past data and an average ratio obtained. In this method the ratios should sum to 4 for quarterly data or 12 for monthly data. If the sum is not correct each ratio must be corrected by dividing it by the average 'average ratio'. The forecast is then given by:

$$y_{T,\,t(J)} = [a_T + b_T \cdot (t - T)] \cdot q_j$$

The values of a_T and b_T are updated as usual. There appears to be no infallible rule as to which method is best in any particular situation and it is necessary to try each on the available data and choose that giving the smallest errors.

Note. q_j is sometimes used as a seasonal index number. Future data may be deseasonalized by dividing by q_j, i.e. deseasonalized demand $= d_i/q_j$.

The data will not always show an annual seasonal pattern but may indicate cycles of a length other than a year. The methods described above may still be applied but the period of the moving average for 'deseasonalizing' the data to calculate the seasonal deviations must now equal the cycle length. In some cases an estimate of the cycle length may be obtained by studying the graph. In other cases it may be necessary to try different values for the period of the moving average and choose that which gives the smoothest trend line. It should be noted that this is a crude method of finding the cycle length and that more sophisticated methods exist if required.

11.5 CHOOSING THE APPROPRIATE MODEL

Having discussed three types of model, there remains the problem of deciding which one to use in a particular situation. In some situations the following process of choice may be useful:

1. Plot historical data on a graph as a time series.
2. Is there apparently a significant cyclic pattern or are the variations reasonably even about a trend line?
3. If there is a cyclic pattern, determine the cycle length and deseasonalize the data. Calculate the average cyclic deviation for each type of period.
4. Determine whether to apply a constant or a linear model to the resulting trend line—perhaps by least squares.
5. In general, use the exponential smoothing method. Determine starting values for a_T, b_T, and $y_{T,t(j)}$ from the historical trend line, preferably leaving a few periods worth of data over which to run in the forecasting system up to 'now'. Use $\beta = 0.8$ for the constant model, $\beta = 0.9$ for the trend model.
6. Start forecasting and update each period.

Other situations will arise in which none of the above methods will be suitable. These may become evident if the one-step ahead errors, e_T, are consistently too large. In these situations it may be necessary to use more sophisticated forecasting techniques. (Reference Brown [3] or Lewis [8].)

11.6 CONTROL OF FORECASTS AND ADAPTIVE SYSTEMS

Frequently the types of forecasting system discussed are operated either manually by relatively unskilled hands or mechanically by a computer. In either case it is necessary to have some warning when the system has gone out of control and is producing wildly erroneous forecasts.

The forecasts are not expected to be perfect and it is possible to calculate confidence limits within which an observation can be expected to fall with a probability of 0·95.

The errors may be plotted on a Statistical Quality Control chart or a Cumulative Sum Chart (Reference [6]) on which warning and action limits are imposed. When the error curve crosses one of these limits the forecasting system is said to have gone out of control and the constants in the equation may need correcting. At this stage the underlying cause of the loss of control should be ascertained since there may have been a significant change in the market environment. Another method for testing whether the system is in control is by the use of tracking signals for details of which the reader is referred to Brown [3], Trigg [4].

Systems can be designed which have a built in correction process, i.e. they adapt themselves to changes in the demand pattern. This usually involves having a varying value for β which is updated each period. References Trigg and Leach [5], Batty [2].

11.7 SUMMARY

Three types of demand model and their appropriate forecasting equations have been discussed.

Constant demand model

$$d_i = a + \varepsilon_i$$

Forecasting equation:

$$y_{T,t} = \beta . y_{T-1,t} + (1 - \beta)d_T$$

where $\beta = 0.8$

Trend model

$$d_i = a + b . i + \varepsilon_i$$

Forecasting equation:

$$y_{T,t} = a_T + b_T . (t - T)$$

where $a_T = a_{T-1} + b_{T-1} + (1 - \beta^2) . e_T$
$b_T = \qquad b_{T-1} + (1 - \beta)^2 . e_T$
$e_T = d_T - y_{T-1,T}$
and $\beta = 0.9$

Seasonal model

Forecasting equation:

$$y_{T,t(j)} = a_T + b_T \cdot (t - T) + p_j$$

where a_T and b_T are updated as in the trend model.

The constant and trend forecasting equations are very effective although it is admitted that the additive and ratio seasonal methods are somewhat crude but also effective. More sophisticated forecasting techniques are available but it is often doubtful whether their use is justified in terms of a cost–benefit analysis. Two books for further reading are Battersby [1] and I.C.I. Monograph No. 2 [7].

EXERCISES

1. Demand has been constant at a level of 9 units per month for some time up to and including month zero. Thereafter it has the following pattern:

i	0	1	2	3	4	5	6	7	8
d_i	9	11	10	7	9	9	11	8	7

Calculate:

(*a*) the four month moving average from month 1 to month 8

(*b*) the exponential smoothing forecast from $i = 1$ to $i = 8$ using $\beta = 0.8$. (This value of β can be seen to smooth out most of the noise.)

(*c*) the exponential smoothing forecast from $i = 1$ to $i = 8$ using $\beta = 0.2$. (This value of β can be seen to give forecasts which are sensitive to demand fluctuations.)

In all three cases plot the forecasts at one period ahead, i.e. plot $y_{T, T+1}$ at $i = T + 1$, and compare with the plots of demand.

2. Assuming a linear trend model, set up a forecasting system using the last 18 months of demand data. Use the first 12 months of that data to set values to the parameters a and b (fit a straight line by eye) and use the remaining 6 months data to run the system in so as to smooth out the effect of minor errors in setting the initial values. $\beta = 0.9$.

i	1	2	3	4	5	6	7	8	9	10	11	12	13	14	15	16	17	18(now)
d_i	20	22	21	23	20	22	$21\frac{1}{4}$	$23\frac{1}{2}$	21	25	$22\frac{1}{4}$	24	$23\frac{1}{4}$	26	$24\frac{1}{4}$	$26\frac{1}{4}$	27	$25\frac{1}{4}$

Use the system to make forecasts for month 19 and for month 23.

3. Assuming a seasonal model, set up a forecasting system using the last 30 months of demand data. Use the first 24 months data to calculate the average seasonal deviations and to set the initial values to the trend line parameters, *a* and *b*. (Fit the line by eye.) Use the remaining 6 months data to run the system in. $\beta = 0.9$.

i	1	2	3	4	5	6	7	8	9	10	11	12	13	14	15
d_t	2	0	0	2	6	6	8	14	10	6	6	4	4	0	4

i	16	77	18	19	20	21	22	23	24	25	26	27	28	29	30
d_t	10	6	10	8	14	18	12	6	10	6	4	8	12	8	12

Use the system to make forecasts for months 31 and 34.

4. The following are the last three years worth of quarterly data on demand for a product *Y*:

Quarter	1	2	3	4	5	6	7	8	9	10	11	12(now)
Demand	30	25	31	26	25	17	22	12	15	16	10	12

Use the data to set up an exponential smoothing forecasting system with a weighting factor equal to 0·9 and forecast demand for the 13th and 14th quarters.

(University of Strathclyde)

5. An ice-cream manufacturer wishes to estimate his takings in Summer 1977 to decide if an increase in production facilities is required. Using the information given below and taking account of the seasonal fluctuation, calculate this value from a line fitted by eye to the moving average trend.

Manufacturers Takings—Thousands of Pounds

	1976	1975	1974
Spring quarter	3	2	2
Summer quarter	8	7	5
Autumn quarter	4	4	3
Winter quarter	3	2	1

(IOS, Part 1)

[*Hint.* Do not use a running-in period since only a once-off forecast is required and not an ongoing forecasting system.]

6.

Hosiery Manufacturers' Sales of Men's Pullovers, Jumpers, Cardigans, etc.

Monthly Average—Millions

Year	Quarters			
	1st quarter	*2nd quarter*	*3rd quarter*	*4th quarter*
1964	1·30	1·23	1·51	1·56
1965	1·27	1·29	1·72	1·66
1966	1·30	1·35	1·73	1·59
1967	1·15	1·18	1·46	1·88
1968	1·33	1·50	1·75	

Calculate the regular seasonal movement of the above series by using the method of moving averages.

(IM, Part 2)

7. Forecast demand for the second quarter of 1978 from the following quarterly demand data:

Year	1st quarter	2nd quarter	3rd quarter	4th quarter
		Quarters		
1974	10	10	9	14
1975	17	15	16	19
1976	22	21	21	24
1977	27	27	28	32

(From University of Strathclyde)

12. *Some Analyses of Market Studies*

12.1 INTRODUCTION

There are essentially two types of study that can be carried out on the market. The first of these is *experimental* in nature and an example would be a test marketing operation. The researcher is able to put varying values to the parameters of the situation and measure their effect on some dependent variable.

The second type of study is non-experimental or *observational*. An example is a market survey. Here the researcher has no control over the variables in the market and he can only measure their current values and then draw inferences about the inter-relationships between them.

The purpose of this chapter is to acquaint the reader with some of the terms used in the analyses of such studies and to give him some indication as to the validity of using such analyses in a given situation.

12.2 EXPERIMENTAL STUDIES

Experiments can be designed to test the effect on sales of different advertising levels, package designs, merchandising displays, prices, etc. It is usually necessary to compare the experimental values with a control which may be taken as the current value of the particular factor, e.g. current packaging or current price.

A specific example would be an experiment to test the effect of three different merchandising displays for salad cream. Each display, *A*, *B*, and *C*, is used in turn in each of five different stores, maybe for two weeks in each store. The sales of salad cream for each two week period in each store are measured by a retail audit and the resulting data is of the form shown in Table 12.1.

Display *A* may be viewed as the control group, i.e. it is the normal display. Displays *B* and *C* are new displays which are being evaluated. It is now necessary to see whether the data provides evidence of a significant improvement on sales due to one or other of the new displays or whether the differences in mean sales levels are due to chance alone.

The data could be analysed by calculating the mean for each display and then testing for significant difference between each pair of means by a *t*-test (*see* Section 6.6) which tests two means in isolation from the others and may show greater differences than really exist across the board. The method of *analysis of variance* is used to test all the means together.

In essence the method analyses the total variance of the data (i.e. variation about the grand mean) by splitting it, in the above example, into two parts: that caused by the effects of different merchandising displays; and that caused by chance or randomness. The variance due to displays is measured by the amount that the individual display means are dispersed about the overall mean and is known as the *between displays* variance. The variance due to chance effects is

TABLE 12.1

Sales of salad cream

Store	Merchandising display			
	A	B	C	
1	16	20	18	
2	15	16	19	
3	16	21	19	
4	15	16	18	
5	16	17	19	
Sum	78	90	93	261
Mean	15·6	18·0	18·6	17·4 (= overall mean)

measured by the amount that the observations are dispersed about their own display means and is known as the *within displays* or *residual* variance. If, as in this particular example, the between displays variance is significantly larger than the residual variance when tested by an *F*-test there are said to be differences between merchandising displays. In other words the total variability in the data cannot be explained merely by the effects of randomness but must be due in part to the different effects of the merchandising displays. The analysis of variance does not go as far as identifying which of the displays are significantly better or worse than the others but the display means may now be tested using a multiple confidence interval test (*see* Davies [4]). In this example it can be seen by inspection that displays *B* and *C* are better than display *A*. In other examples the differences may not be so marked and yet the analysis of variance may still show them to be significant. If the between displays variance had not been significant it would be argued that all the variability in the data was due to chance alone.

In the above example it is assumed that only merchandising displays have an effect on sales and the analysis is said to be carried out on one classification. In another experiment it might be believed that different stores also have different effects on sales and the analysis could be carried out by two classifications, displays and stores. The total variability is then split three ways and assigned to the effects of displays, stores, and chance. Displays and stores variances are then tested against the residual variance, i.e. that due to chance. It is possible to analyse experiments where there are more than two factors which are believed to have an effect on the variable being measured and many standard experimental designs have been developed. These include completely randomized designs, randomized block designs, Latin Squares, and nested or hierarchical designs, (Cox [3], Cochrane and Cox [1], Duckworth [5]). All these designs rest on the assumption of a normal population but an alternative analysis procedure for the one-way classification which does not depend on this assumption is the Kruskal and Wallis test described in Section 7.9.

Because experiments can be controlled and can be designed in a way such that the results reflect only the effect of the factors under study, they should produce more reliable information for decision-making purposes than can observational studies. The implication is then that experimental studies should be carried out in preference to observational studies but there are a number of reasons why this is often infeasible. Firstly, the factor or factors under study may not be under the complete control of the researcher. For instance a manufacturer may wish to find the effect on retail sales of different prices but since he may only recommend prices to the retailer he does not have complete control over the eventual retail prices. Thus the amount of experimentation that he can do is limited. Even if the manufacturer could get the retailer to co-operate by selling at different prices in different periods, it may be that sales would fall off to such an extent that the retailer might decide to call off the experiment before it was properly finished thereby ruining the experiment. Secondly, the experiment may be economically or physically infeasible. Suppose a company wished to speed up its delivery of orders in a certain area by opening up a new warehouse. An experiment to assess the project could involve opening the warehouse and measuring any improvement in customer service. However the experiment is then not an experiment at all but has become the project and could be very expensive if the desired effects were not achieved. Finally since most experiments take a long time to set up and carry out the situation may have changed so that the results when they arrive are no longer relevant to the problem. Also it may be that the time scale for making the decision is much shorter than that necessary for the experiment. For these reasons it is often necessary to carry out observational studies instead.

12.3 OBSERVATIONAL STUDIES

An observational study is one in which the parameters of the situation are left unaltered, i.e. they keep their current values, and observations are made on the various factors. The observations often take the form of a questionnaire-type survey and there is abundant literature on the way that questionnaires should be drawn up and samples selected (References [7], [8], and [11]). The analysis consists of drawing inferences from the observations about the inter-relationships between the various factors. It should be noted that the proposed method of analysis should be borne in mind when designing the survey.

12.3.1 CROSS-CLASSIFICATION

Cross-classification is one of the most common methods of survey analysis. In consumer surveys the consumer may be classified by age, sex, marital status, or socio-economic group. In industrial surveys the customer may be classified by industry, turnover, number employed or capital employed. In either case the company wishes to determine its consumer or customer profile, i.e. the class or combination of classes which makes up the group that buys its product and those which do not. Alternatively it may wish to know which group likes some feature of its product and which does not. Generally 'buy/no buy' is tabulated against customer classification and conclusions on the consumer profile are drawn from the percentages in each cell.

For example suppose a survey is carried out on gramophone record pur-

TABLE 12.2

Percentage of 'pop' record purchasers by sex

	Male	*Female*
'Pop' purchasers	51	63
Non-purchasers	49	37
Total	100	100
No. of respondents	370	330

chasers to determine the profile of purchasers of 'popular' records. A sample of 700 respondents is taken. The resulting data could be classified according to sex as shown in Table 12.2 and a conclusion drawn that females are more likely purchasers of 'pop' records than males. Alternatively the data could be classified according to age group as in Table 12.3 and a conclusion drawn that the younger

TABLE 12.3

Percentage of 'pop' record purchasers by age group

	Below 20	*AGE GROUP* *20–30*	*Above 30*
'Pop' purchasers	74	57	37
Non-purchasers	26	43	63
Total	100	100	100
No. of respondents	230	260	210

people tend to be 'pop' purchasers. The data has now been analysed by two factors in succession but has revealed nothing about the inter-relation between sex and age group. This type of information can be obtained only by a simultaneous analysis in which, say, each sex is subdivided into age groups as in

TABLE 12.4

Percentage of 'pop' record purchasers by sex and age group

	Male			Female		
	<20	20–30	>30	<20	20–30	>30
'Pop' purchasers	65	55	35	80	60	40
Non-purchasers	35	45	65	20	40	60
Total	100	100	100	100	100	100
No. of respondents	100	150	120	130	110	90

Table 12.4. From this table it can be concluded that young females are the main purchasers but the profile is the overall distribution of purchasers.

In practice there is something of an art in the way in which classes are tabulated. The independent variables such as sex and age group should only be analysed successively if it is a reasonable assumption that there are no relationships between them. If this assumption is suspected to be invalid it is necessary to analyse all the variables simultaneously. However simultaneous analysis leads to classification under many subheadings and hence large data requirements to obtain reasonable estimates of the percentages in each cell. Only very few independent variables can be handled by this method since data handling and interpretation of results rapidly become complex. There is no guarantee that all the relevant variables will have been included in the analysis because the method can handle so few with comfort anyway.

The method of cross-classification is probably in common use because the tabulations may be interpreted without sophisticated statistical analysis although it should be appreciated that each tabulation is theoretically just one sample out of a large number of possible samples. There are therefore likely to be estimation errors but confidence in the estimates may be increased by increasing the sample size—at a cost.

12.3.2 MULTIPLE REGRESSION

Multiple regression was briefly mentioned in Chapter 10 (*see* page 190) and is a technique for relating two or more independent variables to one dependent variable. It is sometimes applicable in situations where the dependent variable is measurable, e.g. sales quantity or turnover, market share, etc.

An equation of the form:

$$Y = b_0 + b_1x_1 + b_2x_2 + \ldots + b_px_p$$

where Y = dependent variable (e.g. sales turnover)

x_i = ith independent variable (e.g. advertising expenditure or market share)

b_i = constant associated with x_i

is fitted to the data by the method of Least Squares in the same manner as for a linear regression equation. The calculations would be difficult to handle manually and therefore they are usually carried out on a computer using a standard program or package.

It may be that there is data available on many independent variables that are believed to have some effect on the dependent variable. However it might be possible to explain most of the variation in the Y values by a regression equation containing only a few of the available dependent variables, i.e. some of the x variables may have a more significant effect on Y than some of the others. Most packages produce a regression equation for Y only in terms of these significant x variables. A *multiple correlation coefficient* (R) analogous to the linear correlation coefficient (r) is also calculated and its square (R^2) gives the proportion of the variation in y values explained by the fitted regression equation. There is usually no restriction on whether x variables are correlated to one another or not but if they are the final regression equation should be interpreted with care since the effects of the omitted correlated variables may have been masked by those variables included in the equation. No such problem arises if all the x variables are uncorrelated and then the final regression equation only contains the most significant independent variables. It is assumed for multiple regression that the relations between all the variables are linear.

Multiple regression may be used only as a means of identifying the most significant variables or factors in a situation, i.e. its use may be similar to that of cross-classification. This use is most common in market research problems where the researcher wishes to identify those factors which bear on the buy/no buy decision. In situations where future values of the x variables are known or can be estimated multiple regression may be used as a forecasting technique. The whole topic of regression is covered in depth by Smillie [10].

12.3.3 DISCRIMINANT ANALYSIS

When the dependent variable cannot be measured on a continuous scale and can only have values which correspond to categories such as a customer buying or not buying, being a heavy or light user, etc., the significant independent variables may be identified by discriminant analysis. The analysis may then be used to predict the probabilities of a particular customer falling into the different categories given that his values of the independent variables are known.

For example customers may base their 'buy/no buy' decisions for an industrial dyestuff on the different properties that the dyestuff exhibits. Such properties could include the 'fastness' of the colour and the ease of absorption of the dye-

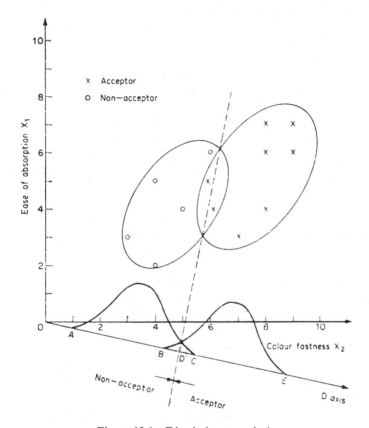

Figure 12.1 Discriminant analysis

stuff on a range of fabric types. Suppose the manufacturer produces a new formula dyestuff and carries out a survey of his customers to determine whether the new formula is acceptable to them. They are asked to give a rating in the range 0 (poor) to 10 (good) for each of the properties of the product and also indicate whether the product is acceptable or not. The ratings can be shown on graphs similar to that in Figure 12.1.

It is possible to fit a linear discriminant function to the rating data:

$$D_i = b_1 X_{1i} + b_2 X_{2i}$$

where: X_{1i} and X_{2i} are the ratings of the ith customer for the two properties;

b_1 and b_2 are estimated constants;

D_i is an estimated combined score for the ith customer.

More terms $b_j X_{ji}$ would be added to the function if more properties were to be considered. In Figure 12.1 ellipses have been drawn around each of the acceptable and unacceptable groups and they are seen to intersect. A line is drawn through the intersections and cuts the D-axis perpendicularly at D'. The two distributions on the D-axis are the distributions of the D_is calculated from the discriminant function for both the acceptable and unacceptable groups, i.e. D_i is related to the plots of X_1 and X_2 through the discriminant function. If a value for D is calculated from the function for a particular customer and if it falls on the D-axis within the shaded area, there is a finite probability that he finds the product either acceptable or unacceptable. The point D' is placed so as to maximize the probability of classifying a customer correctly as an acceptor or not and is the critical level of D which discriminates between acceptors and non-acceptors. If his D-value lies in the sections AB or CE there is a very high probability of his being classified correctly.

If three properties were considered the graph would be drawn in three dimensions and the ellipses would become ellipsoids (i.e. egg-shaped). However the two D_i distributions would still have the same form. The larger the shaded area of overlap of these distributions the less discriminating are the properties with respect to acceptability. The *relative* sizes of the parameters, b_j, are an indication of the relative importance or discriminating power of each property or attribute. This is probably the most valuable result of the analysis.

12.3.4 FACTOR ANALYSIS

Frequently there are a large number of independent variables which appear to affect the dependent variable but which would be very time-consuming and complex to analyse. It is therefore useful to be able to reduce this large set of variables to a smaller set on which regression or discriminant analysis may be carried out. Factor analysis is concerned with this reduction and in such a way that the resulting small set adequately represents the large set. It either identifies the significant variables in the large set or produces significant combinations of some of the original variables.

Methods of factor analysis were developed early in the century mainly for use by psychologists. Factor analysis is a collective term for a number of techniques which analyse the inter-correlations within a set of variables. Principal components analysis is one such technique.

Examples of the use of factor analysis in marketing studies are quoted in Frank and Green [6]. A particular study of the variation in television ownership used a factor analysis of twenty-seven variables to reduce that number to ten new measures which were then used in a multiple regression analysis.

12.4 SUMMARY

Studies of the market may be either experimental or observational in nature. Experimental studies should be carefully designed and controlled in order to avoid undesirable bias and so that the results only reflect controlled changes in the values of the variables under study. These conditions are often difficult to meet in practice. The results are often analysed by analysis of variance techniques.

Observational studies are often of the survey type which should also be carefully designed. The researcher has no control over the values of the variables under study and the results are likely to be contaminated by the effects of chance and other variables. He has to identify the true inter-relationships between variables from this contaminated data. Conceptually the simplest form of analysis is that of cross-classification although data handling and interpretation becomes difficult with more than a few variables.

More powerful techniques are those of multiple regression and discriminant analysis. The former is used where the dependent variable may be measured (e.g. sales volume) and the latter is used where the dependent variable is categorical (e.g. buy or not buy). If a large number of variables can be included in either the regression equation or the discriminant function it may be simpler to reduce that number first to a few significant factors by factor analysis.

These and other multivariate techniques tend to require large quantities of data which makes manual analysis time consuming and tedious. Computer packages or programs are usually available for them but interpretation of the output results may be difficult and it is advisable to call in a statistician to advise on such studies.

Further reading: Cooley and Lohnes [2] and Ralston and Wilf [9].

Appendix A

(I) PROOF OF EQUIVALENCE OF FORMULAE 2.3 AND 2.5

Formula 2.3 is:

$$s^2 = \frac{\sum_{i=1}^{n} (x_i - m)^2}{n} \quad \text{(for ungrouped data)} \qquad [2.3]$$

Expansion of $(x_i - m)^2$ is $x_i^2 - 2x_im + m^2$ and therefore:

$$\sum_{i=1}^{n} \frac{(x_i - m)^2}{n} = \frac{1}{n}\left[\sum_{i=1}^{n} (x_i^2 - 2x_im + m^2)\right]$$

which is equal to:

$$\frac{\sum_{i=1}^{n} x_i^2}{n} - \frac{2m \sum_{i=1}^{n} x_i}{n} + \frac{\sum_{i=1}^{n} m^2}{n}$$

$$= \frac{\sum_{i=1}^{n} x_i^2}{n} - 2m^2 + m^2$$

$$= \frac{\sum_{i=1}^{n} x_i^2}{n} - m^2 \qquad [2.5]$$

since $\dfrac{\sum_{i=1}^{n} x_i}{n} = m$ and $\sum_{i=1}^{n} m^2 = nm^2$

$$\therefore \frac{\sum_{i=1}^{n} (x_i - m)^2}{n} = \frac{\sum_{i=1}^{n} x_i^2}{n} - m^2$$

A similar proof can be constructed for the formulae for grouped data, namely Formulae 2.4 and 2.6.

(2) PROOF OF EQUIVALENCE OF FORMULAE 2.1 AND 2.7

Formula 2.1 is:

$$m = \frac{\sum\limits_{i=1}^{n} x_i}{n} \qquad [2.1]$$

Introducing a new variable $u_i = \dfrac{x_i - a}{c}$

we get:

$$x_i = cu_i + a$$

$$\therefore m = \frac{\sum\limits_{i=1}^{n} (cu_i + a)}{n}$$

$$= \frac{c \sum\limits_{i=1}^{n} u_i + \sum\limits_{i=1}^{n} a}{n}$$

$$\therefore m = c\bar{u} + a \qquad [2.7]$$

since $\dfrac{\sum\limits_{i=1}^{n} u_i}{n} = \bar{u}$ and $\dfrac{\sum\limits_{i=1}^{n} a}{n} = a$

(3) PROOF OF EQUIVALENCE OF FORMULA 2.3 AND FORMULA 2.8

Formula 2.3 is:

$$s^2 = \frac{\sum\limits_{i=1}^{n} (x_i - m)^2}{n} \qquad [2.3]$$

$$= \frac{\sum\limits_{i=1}^{n} x_i^2}{n} - m^2$$

Defining a new variable as $u_i = \dfrac{x_i - a}{c}$ we get: $x_i = cu_i + a$

Now $m = c\bar{u} + a$

$$\therefore s^2 = \frac{\sum\limits_{i=1}^{n} (cu_i + a - c\bar{u} - a)^2}{n}$$

$$= \frac{c^2 \sum\limits_{i=1}^{n} (u_i - \bar{u})^2}{n}$$

Now

$$\frac{\sum_{i=1}^{n} (u_i - \bar{u})^2}{n} = \frac{\sum_{i=1}^{n} u_i^2}{n} - \bar{u}^2$$

$$\therefore s^2 = c^2 \left[\frac{\sum_{i=1}^{n} u_i^2}{n} - \bar{u}^2 \right]$$

A similar derivation can be performed to prove the equivalence of Formulae 2.4 and 2.9 which are appropriate for grouped data.

Appendix B

PROPERTIES OF ESTIMATORS

In Section 4.5 the sample mean and $n/(n-1) \times$ sample variance are cited as 'best' estimators of population mean and variance. The reader is asked to interpret 'best' estimator as the estimator which would be preferred in the great majority of cases and which would be acceptable in other cases. The reasons for this preference for the sample mean from a normal distribution are as follows:

1. The expected value of the sample mean is the population mean. The sample mean is therefore said to be an *unbiased* estimator of the population mean.

2. The variance of the sample mean is σ^2/n and therefore as sample size increases the variance of the sample mean decreases. This implies that as sample size increases individual sample means can be considered as clustering more and more closely around the population mean value. The sample mean is therefore said to be a *consistent* estimator of the population mean.

3. It can be shown that no unbiased estimator of the population mean can have a smaller variance than σ^2/n. Hence the sample mean is said to be a *minimum variance* estimator of the population mean.

These properties can be recognized intuitively as desirable properties of an estimator. Similar statements can be made about sample variance with the important difference that the factor $n/(n-1)$ must be introduced to give an unbiased estimator.

Appendix C

Optimal sample size for quadratic error costs

$$T_c = C_1 + nC_2 + \frac{k\sigma^2}{n}$$

The only unknown value in this equation is that of n. The value of n which minimizes T_c is found by differentiating T_c with respect to n, equating the derivative to zero and solving for n:

$$\frac{d(T_c)}{dn} = C_2 - \frac{k\sigma^2}{n^2} = 0$$

$$\therefore n^2 = \frac{k\sigma^2}{C_2}$$

$$n = \sqrt{\frac{k\sigma^2}{C_2}}$$

Appendix D

A population of N items with mean μ can be divided into two strata A and B containing respectively N_A items with variance σ_A^2 and N_B items of variance σ_B^2. A sample of overall size n is to be partitioned between A and B so that the variance of the estimate of μ is minimized.

Suppose that the optimal partition is k items drawn from A and $n - k$ from B.

k items give a mean m_A with sampling variance $\dfrac{\sigma_A^2}{k}$

$n - k$ items give a mean m_B with sampling variance $\dfrac{\sigma_B^2}{n - k}$

Estimate of $\mu = \dfrac{N_A}{N} \times m_A + \dfrac{N_B}{N} \times m_B$

Because variances may be added and because variance $(Cx) = C^2$ (variance (x)) where C is a constant —

$$\text{Variance of estimate of } \mu = \left(\frac{N_A}{N}\right)^2 \frac{\sigma_A^2}{k} + \left(\frac{N_B}{N}\right)^2 \frac{\sigma_B^2}{n - k}$$

By differentiation the value of k which minimizes this variance is given by:

$$-\left(\frac{N_A}{N}\right)^2 \frac{\sigma_A^2}{k^2} + \left(\frac{N_B}{N}\right)^2 \frac{\sigma_B^2}{(n - k)^2} = 0$$

This gives:

$$\frac{k^2}{(n - k)^2} = \frac{N_A^2}{N^2} \sigma_A^2 \frac{N^2}{N_B^2 \sigma_B^2} = \frac{(N_A \sigma_A)^2}{(N_B \sigma_B)^2}$$

$$\frac{k}{n - k} = \frac{N_A \sigma_A}{N_B \sigma_B}$$

Hence the sample is divided in proportion to the product of the number of items and standard deviation of each stratum. This generalizes for more than two strata to the statement that the number of items to be drawn from stratum A is:

$$n_A = \left(\frac{N_A \sigma_A}{N_A \sigma_A + N_B \sigma_B + N_C \sigma_C + \ldots}\right) \times n$$

Appendix E

When the 'F' test shows a significant difference between the two estimates of population variance the following procedure may be adopted to compare the means of two samples.

If the two samples are characterized by:

$$n_1, s_1^2, m_1 \quad \text{and} \quad n_2, s_2^2, m_2$$

$$\hat{\sigma}_1^2 = \frac{n_1 s_1^2}{n_1 - 1} \quad \sigma_2^2 = \frac{n_2 s_2^2}{n_2 - 1}$$

Calculate:

$$t = \frac{m_1 - m_2}{\sqrt{\dfrac{\hat{\sigma}_1^2}{n_1} + \dfrac{\hat{\sigma}_2^2}{n_2}}}$$

and look up 't' table with:

$$\frac{\left(\dfrac{\hat{\sigma}_1^2}{n_1} + \dfrac{\hat{\sigma}_2^2}{n_2}\right)^2}{\left(\left(\dfrac{\hat{\sigma}_1^2}{n_1}\right)^2 \times \dfrac{1}{n_1 + 1} + \left(\dfrac{\hat{\sigma}_2^2}{n_2}\right)^2 \times \dfrac{1}{n_2 + 1}\right)} \quad \text{degrees of freedom}$$

This last value is not necessarily an integer so interpolation in the tables is necessary.

Appendix F

DERIVATIONS OF FORECASTING EQUATIONS

1. EQUATION [11.2]

Starting with equation [11.1]

$$y_{T,t} = \frac{1}{N} \sum_{i=T-N+1}^{T} d_i \qquad \text{(i)}$$

Similarly the forecast at the previous period, $T - 1$, is:

$$y_{T-1,t} = \frac{1}{N} \sum_{i=T-N}^{T-1} d_i \qquad \text{(ii)}$$

Subtracting (ii) from (i) gives:

$$y_{T,t} - y_{T-1,t} = \frac{1}{N} \sum_{i=T-N+1}^{T} d_i - \frac{1}{N} \sum_{i=T-N}^{T-1} d_i$$

$$= \frac{1}{N}(d_T - d_{T-N})$$

$$y_{T,t} = y_{T-1,t} + \frac{1}{N}(d_T - d_{T-N}) \qquad [11.2]$$

2. EXPONENTIAL SMOOTHING EQUIVALENT OF EQUATION [11.1]

Starting with equation [11.3]:

$$y_{T,t} = \beta \cdot y_{T-1,t} + (1 - \beta) \cdot d_T \qquad \text{(i)}$$

Similarly the forecast at the previous period, $T - 1$, is:

$$y_{T-1,t} = \beta \cdot y_{T-2,t} + (1 - \beta) \cdot d_{T-1} \qquad \text{(ii)}$$

Substituting (ii) in (i) gives:

$$y_{T,t} = \beta^2 \cdot y_{T-2,t} + (1 - \beta) \cdot (d_T + \beta \cdot d_{T-1}) \qquad \text{(iii)}$$

The forecast made at period $(T - 2)$ is:

$$y_{T-2,t} = \beta \cdot y_{T-3,t} + (1 - \beta) \cdot d_{T-2} \qquad \text{(iv)}$$

Substituting (iv) in (iii) gives:

$$y_{T,t} = \beta^3 \cdot y_{T-3,t} + (1 - \beta) \cdot (d_T + \beta \cdot d_{T-1} + \beta^2 \cdot d_{T-2}) \qquad \text{(v)}$$

Equation (v) may be generalized to give the forecast made now, T, in terms of the forecast made r periods ago, $T - r$, i.e.:

$$y_{T,t} = \beta^r \cdot y_{T-r,t} + (1 - \beta) \cdot (d_T + \beta \cdot d_{T-1} + \beta^2 \cdot d_{T-2} + \ldots + \beta^r \cdot d_{T-r})$$

$$= \beta^r \cdot y_{T-r,t} + (1 - \beta) \cdot \sum_{i=T-r}^{T} \beta^{T-i} \cdot d_i \ldots \qquad \text{(vi)}$$

It can now be seen that an observation r periods old has a weight of $\beta^r(1 - \beta)$.

As r becomes large, i.e. many old observations are included, the term $\beta^r \cdot y_{T-r,t}$ becomes negligible and may be omitted from (vi) leaving

$$y_{T,t} = (1 - \beta) \cdot \sum_{i=T-r}^{T} \beta^{T-i} \cdot d_i \qquad \text{(vii)}$$

Equation (vii) is approximately the weighted sum of the last r observations divided by the sum of the weights

since $\quad 1 + \beta + \beta^2 + \ldots + \beta^r = \dfrac{1}{(1 - \beta)} \quad$ as r tends to infinity.

Thus equation (vii) is equivalent to the moving average Equation [11.1].

Tables of Statistical Distributions

Tables T2, T3, T4a, T4b, and T5 are taken from Tables III, IV, V, and XXXIII of Fisher and Yates: *Statistical Tables for Biological, Agricultural and Medical Research*, (Oliver & Boyd, 1957), by permission of the authors and publishers.

TABLE T.1
Areas under the standard normal curve

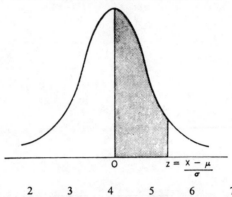

$$z = \frac{x - \mu}{\sigma}$$

z	0	1	2	3	4	5	6	7	8	9
0·0	0·0000	0·0040	0·0080	0·0120	0·0160	0·0199	0·0239	0·0279	0·0319	0·0359
0·1	0·0398	0·0438	0·0478	0·0517	0·0557	0·0596	0·0636	0·0675	0·0714	0·0754
0·2	0·0793	0·0832	0·0871	0·0910	0·0948	0·0987	0·1026	0·1064	0·1103	0·1141
0·3	0·1179	0·1217	0·1255	0·1293	0·1331	0·1368	0·1406	0·1443	0·1480	0·1517
0·4	0·1554	0·1591	0·1628	0·1664	0·1700	0·1736	0·1772	0·1808	0·1844	0·1879
0·5	0·1915	0·1950	0·1985	0·2019	0·2054	0·2088	0·2123	0·2157	0·2190	0·2224
0·6	0·2258	0·2291	0·2324	0·2357	0·2389	0·2422	0·2454	0·2486	0·2518	0·2549
0·7	0·2580	0·2612	0·2642	0·2673	0·2704	0·2734	0·2764	0·2794	0·2823	0·2852
0·8	0·2881	0·2910	0·2939	0·2967	0·2996	0·3023	0·3051	0·3078	0·3106	0·3133
0·9	0·3159	0·3186	0·3212	0·3238	0·3264	0·3289	0·3315	0·3340	0·3365	0·3389
1·0	0·3413	0·3438	0·3461	0·3485	0·3508	0·3531	0·3554	0·3577	0·3599	0·3621
1·1	0·3643	0·3665	0·3686	0·3708	0·3729	0·3749	0·3770	0·3790	0·3810	0·3830
1·2	0·3849	0·3869	0·3888	0·3907	0·3925	0·3944	0·3962	0·3980	0·3997	0·4015
1·3	0·4032	0·4049	0·4066	0·4082	0·4099	0·4115	0·4131	0·4147	0·4162	0·4177
1·4	0·4192	0·4207	0·4222	0·4236	0·4251	0·4265	0·4279	0·4292	0·4306	0·4319
1·5	0·4332	0·4345	0·4357	0·4370	0·4382	0·4394	0·4406	0·4418	0·4429	0·4441
1·6	0·4452	0·4463	0·4474	0·4484	0·4495	0·4505	0·4515	0·4525	0·4535	0·4545
1·7	0·4554	0·4564	0·4573	0·4582	0·4591	0·4599	0·4608	0·4616	0·4625	0·4633
1·8	0·4641	0·4649	0·4656	0·4664	0·4671	0·4678	0·4686	0·4693	0·4699	0·4706
1·9	0·4713	0·4719	0·4726	0·4732	0·4738	0·4744	0·4750	0·4756	0·4761	0·4767
2·0	0·4772	0·4778	0·4783	0·4788	0·4793	0·4798	0·4803	0·4808	0·4812	0·4817
2·1	0·4821	0·4826	0·4830	0·4843	0·4838	0·4842	0·4846	0·4850	0·4854	0·4857
2·2	0·4861	0·4864	0·4868	0·4871	0·4875	0·4878	0·4881	0·4884	0·4887	0·4890
2·3	0·4893	0·4896	0·4898	0·4901	0·4904	0·4906	0·4909	0·4911	0·4913	0·4916
2·4	0·4918	0·4920	0·4922	0·4925	0·4927	0·4929	0·4931	0·4932	0·4934	0·4936
2·5	0·4938	0·4940	0·4941	0·4943	0·4945	0·4946	0·4948	0·4949	0·4951	0·4952
2·6	0·4953	0·4955	0·4956	0·4957	0·4959	0·4960	0·4961	0·4962	0·4963	0·4964
2·7	0·4965	0·4966	0·4967	0·4968	0·4969	0·4970	0·4971	0·4972	0·4973	0·4974
2·8	0·4974	0·4975	0·4976	0·4977	0·4977	0·4978	0·4979	0·4979	0·4980	0·4981
2·9	0·4981	0·4982	0·4982	0·4983	0·4984	0·4984	0·4985	0·4985	0·4986	0·4986
3·0	0·4987	0·4987	0·4987	0·4988	0·4988	0·4989	0·4889	0·4989	0·4990	0·4990
3·1	0·4990	0·4991	0·4991	0·4991	0·4992	0·4992	0·4992	0·4992	0·4993	0·4993
3·2	0·4993	0·4993	0·4994	0·4994	0·4994	0·4994	0·4994	0·4995	0·4995	0·4995
3·3	0·4995	0·4995	0·4995	0·4996	0·4996	0·4996	0·4996	0·4996	0·4996	0·4997
3·4	0·4997	0·4997	0·4997	0·4997	0·4997	0·4997	0·4997	0·4997	0·4997	0·4998
3·5	0·4998	0·4998	0·4998	0·4998	0·4998	0·4998	0·4998	0·4998	0·4998	0·4998

236

Distribution of t with v degrees of freedom

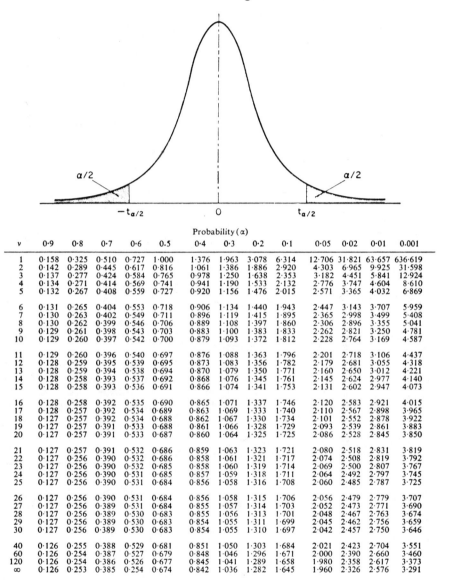

Probability (α)

v	0·9	0·8	0·7	0·6	0·5	0·4	0·3	0·2	0·1	0·05	0·02	0·01	0·001
1	0·158	0·325	0·510	0·727	1·000	1·376	1·963	3·078	6·314	12·706	31·821	63·657	636·619
2	0·142	0·289	0·445	0·617	0·816	1·061	1·386	1·886	2·920	4·303	6·965	9·925	31·598
3	0·137	0·277	0·424	0·584	0·765	0·978	1·250	1·638	2·353	3·182	4·451	5·841	12·924
4	0·134	0·271	0·414	0·569	0·741	0·941	1·190	1·533	2·132	2·776	3·747	4·604	8·610
5	0·132	0·267	0·408	0·559	0·727	0·920	1·156	1·476	2·015	2·571	3·365	4·032	6·869
6	0·131	0·265	0·404	0·553	0·718	0·906	1·134	1·440	1·943	2·447	3·143	3·707	5·959
7	0·130	0·263	0·402	0·549	0·711	0·896	1·119	1·415	1·895	2·365	2·998	3·499	5·408
8	0·130	0·262	0·399	0·546	0·706	0·889	1·108	1·397	1·860	2·306	2·896	3·355	5·041
9	0·129	0·261	0·398	0·543	0·703	0·883	1·100	1·383	1·833	2·262	2·821	3·250	4·781
10	0·129	0·260	0·397	0·542	0·700	0·879	1·093	1·372	1·812	2·228	2·764	3·169	4·587
11	0·129	0·260	0·396	0·540	0·697	0·876	1·088	1·363	1·796	2·201	2·718	3·106	4·437
12	0·128	0·259	0·395	0·539	0·695	0·873	1·083	1·356	1·782	2·179	2·681	3·055	4·318
13	0·128	0·259	0·394	0·538	0·694	0·870	1·079	1·350	1·771	2·160	2·650	3·012	4·221
14	0·128	0·258	0·393	0·537	0·692	0·868	1·076	1·345	1·761	2·145	2·624	2·977	4·140
15	0·128	0·258	0·393	0·536	0·691	0·866	1·074	1·341	1·753	2·131	2·602	2·947	4·073
16	0·128	0·258	0·392	0·535	0·690	0·865	1·071	1·337	1·746	2·120	2·583	2·921	4·015
17	0·128	0·257	0·392	0·534	0·689	0·863	1·069	1·333	1·740	2·110	2·567	2·898	3·965
18	0·127	0·257	0·392	0·534	0·688	0·862	1·067	1·330	1·734	2·101	2·552	2·878	3·922
19	0·127	0·257	0·391	0·533	0·688	0·861	1·066	1·328	1·729	2·093	2·539	2·861	3·883
20	0·127	0·257	0·391	0·533	0·687	0·860	1·064	1·325	1·725	2·086	2·528	2·845	3·850
21	0·127	0·257	0·391	0·532	0·686	0·859	1·063	1·323	1·721	2·080	2·518	2·831	3·819
22	0·127	0·256	0·390	0·532	0·686	0·858	1·061	1·321	1·717	2·074	2·508	2·819	3·792
23	0·127	0·256	0·390	0·532	0·685	0·858	1·060	1·319	1·714	2·069	2·500	2·807	3·767
24	0·127	0·256	0·390	0·531	0·685	0·857	1·059	1·318	1·711	2·064	2·492	2·797	3·745
25	0·127	0·256	0·390	0·531	0·684	0·856	1·058	1·316	1·708	2·060	2·485	2·787	3·725
26	0·127	0·256	0·390	0·531	0·684	0·856	1·058	1·315	1·706	2·056	2·479	2·779	3·707
27	0·127	0·256	0·389	0·531	0·684	0·855	1·057	1·314	1·703	2·052	2·473	2·771	3·690
28	0·127	0·256	0·389	0·530	0·683	0·855	1·056	1·313	1·701	2·048	2·467	2·763	3·674
29	0·127	0·256	0·389	0·530	0·683	0·854	1·055	1·311	1·699	2·045	2·462	2·756	3·659
30	0·127	0·256	0·389	0·530	0·683	0·854	1·055	1·310	1·697	2·042	2·457	2·750	3·646
40	0·126	0·255	0·388	0·529	0·681	0·851	1·050	1·303	1·684	2·021	2·423	2·704	3·551
60	0·126	0·254	0·387	0·527	0·679	0·848	1·046	1·296	1·671	2·000	2·390	2·660	3·460
120	0·126	0·254	0·386	0·526	0·677	0·845	1·041	1·289	1·658	1·980	2·358	2·617	3·373
∞	0·126	0·253	0·385	0·254	0·674	0·842	1·036	1·282	1·645	1·960	2·326	2·576	3·291

TABLE T.3

Distribution of χ^2 with ν degrees of freedom

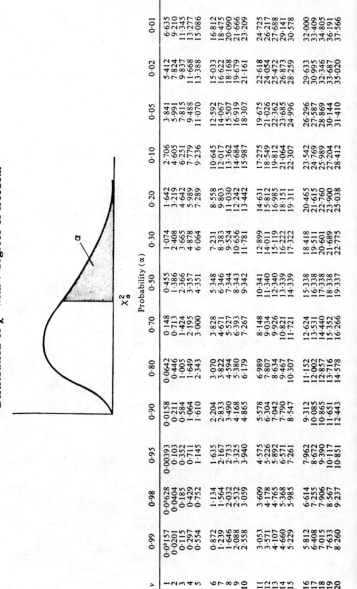

χ^2_α

ν	\multicolumn{14}{c}{Probability (α)}
	0·99	0·98	0·95	0·90	0·80	0·70	0·50	0·30	0·20	0·10	0·05	0·02	0·01	0·001
1	0·0¹157	0·0⁶628	0·0³393	0·0¹158	0·0642	0·148	0·455	1·074	1·642	2·706	3·841	5·412	6·635	10·827
2	0·0201	0·0404	0·103	0·211	0·446	0·713	1·386	2·408	3·219	4·605	5·991	7·824	9·210	13·815
3	0·115	0·185	0·352	0·584	1·005	1·424	2·366	3·665	4·642	6·251	7·815	9·837	11·345	16·266
4	0·297	0·429	0·711	1·064	1·649	2·195	3·357	4·878	5·989	7·779	9·488	11·668	13·277	18·467
5	0·554	0·752	1·145	1·610	2·343	3·000	4·351	6·064	7·289	9·236	11·070	13·388	15·086	20·515
6	0·872	1·134	1·635	2·204	3·070	3·828	5·348	7·231	8·558	10·645	12·592	15·033	16·812	22·457
7	1·239	1·564	2·167	2·833	3·822	4·671	6·346	8·383	9·803	12·017	14·067	16·622	18·475	24·322
8	1·646	2·032	2·733	3·490	4·594	5·527	7·344	9·524	11·030	13·362	15·507	18·168	20·090	26·125
9	2·088	2·532	3·325	4·168	5·380	6·393	8·343	10·656	12·242	14·684	16·919	19·679	21·666	27·877
10	2·558	3·059	3·940	4·865	6·179	7·267	9·342	11·781	13·442	15·987	18·307	21·161	23·209	29·588
11	3·053	3·609	4·575	5·578	6·989	8·148	10·341	12·899	14·631	17·275	19·675	22·618	24·725	31·264
12	3·571	4·178	5·226	6·304	7·807	9·034	11·340	14·011	15·812	18·549	21·026	24·054	26·217	32·909
13	4·107	4·765	5·892	7·042	8·634	9·926	12·340	15·119	16·985	19·812	22·362	25·472	27·688	34·528
14	4·660	5·368	6·571	7·790	9·467	10·821	13·339	16·222	18·151	21·064	23·685	26·873	29·141	36·123
15	5·229	5·985	7·261	8·547	10·307	11·721	14·339	17·322	19·311	22·307	24·996	28·259	30·578	37·697
16	5·812	6·614	7·962	9·312	11·152	12·624	15·338	18·418	20·465	23·542	26·296	29·633	32·000	39·252
17	6·408	7·255	8·672	10·085	12·002	13·531	16·338	19·511	21·615	24·769	27·587	30·995	33·409	40·790
18	7·015	7·906	9·390	10·865	12·857	14·440	17·338	20·601	22·760	25·989	28·869	32·346	34·805	42·312
19	7·633	8·567	10·117	11·651	13·716	15·352	18·338	21·689	23·900	27·204	30·144	33·687	36·191	43·820
20	8·260	9·237	10·851	12·443	14·578	16·266	19·337	22·775	25·038	28·412	31·410	35·020	37·566	45·315

21	8·897	9·915	11·591	13·240	15·445	17·182	20·337	23·858	26·171	29·615	32·671	36·343	38·932	46·797
22	9·542	10·600	12·338	14·041	16·314	18·101	21·337	24·939	27·301	30·813	33·924	37·659	40·289	48·268
23	10·196	11·293	13·091	14·848	17·187	19·021	22·337	26·018	28·429	32·007	35·172	38·968	41·638	49·728
24	10·856	11·992	13·848	15·659	18·062	19·943	23·337	27·096	29·553	33·196	36·415	40·270	42·980	51·179
25	11·524	12·697	14·611	16·473	18·940	20·867	24·337	28·172	30·675	34·382	37·652	41·566	44·314	52·620
26	12·198	13·409	15·379	17·292	19·820	21·792	25·336	29·246	31·795	35·563	38·885	42·856	45·642	54·052
27	12·879	14·125	16·151	18·114	20·703	22·719	26·336	30·319	32·912	36·741	40·113	44·140	46·963	55·476
28	13·565	14·847	16·928	18·939	21·588	23·647	27·336	31·391	34·027	37·916	41·337	45·419	48·278	56·893
29	14·256	15·574	17·708	19·768	22·475	24·577	28·336	32·461	35·139	39·087	42·557	46·693	49·588	58·302
30	14·953	16·306	18·493	20·599	23·364	25·508	29·336	33·530	36·250	40·256	43·773	47·962	50·892	59·703
32	16·362	17·783	20·072	22·271	25·148	27·373	31·336	35·665	38·466	42·585	46·194	50·487	53·486	62·487
34	17·789	19·275	21·664	23·952	26·938	29·242	33·336	37·795	40·676	44·903	48·602	52·995	56·061	65·247
36	19·233	20·783	23·269	25·643	28·735	31·115	35·336	39·922	42·879	47·212	50·999	55·489	58·619	67·985
38	20·691	22·304	24·884	27·343	30·537	32·992	37·335	42·045	45·076	49·513	53·384	57·969	61·162	70·703
40	22·164	23·838	26·509	29·051	32·345	34·872	39·335	44·165	47·269	51·805	55·759	60·436	63·691	73·402
42	23·650	25·383	28·144	30·765	34·157	36·755	41·335	46·282	49·456	54·090	58·124	62·892	66·206	76·084
44	25·148	26·939	29·787	32·487	35·974	38·641	43·335	48·396	51·639	56·369	60·481	65·337	68·710	78·750
46	26·657	28·504	31·439	34·215	37·795	40·259	45·335	50·507	53·818	58·641	62·830	67·771	71·201	81·400
48	28·177	30·080	33·098	35·949	39·621	42·420	47·335	52·616	55·993	60·907	65·171	70·197	73·683	84·037
50	29·707	31·664	34·764	37·689	41·449	44·313	49·335	54·723	58·164	63·167	67·505	72·613	76·154	86·661
52	31·246	33·256	36·437	39·433	43·281	46·209	51·335	56·827	60·332	65·422	69·832	75·021	78·616	89·272
54	32·793	34·856	38·116	41·183	45·183	48·106	53·335	58·930	62·496	67·673	72·153	77·422	81·069	91·872
56	34·350	36·464	39·801	42·937	46·955	50·005	55·335	61·031	64·658	69·919	74·468	79·815	83·513	94·461
58	35·913	38·078	41·492	44·696	48·797	51·906	57·335	63·129	66·816	72·160	76·778	82·201	85·950	97·039
60	37·485	39·699	43·188	46·459	50·641	53·809	59·335	65·227	68·972	74·397	79·082	84·580	88·379	99·607
62	39·063	41·327	44·889	48·226	52·487	55·714	61·335	67·322	71·125	76·630	81·381	86·953	90·802	102·166
64	40·649	42·960	46·595	49·996	54·336	57·620	63·335	69·416	73·276	78·860	83·675	89·320	93·217	104·716
66	42·240	44·599	48·305	51·770	56·188	59·527	65·335	71·508	74·424	81·085	85·965	91·681	95·626	107·258
68	43·838	46·244	50·020	53·548	58·042	61·436	67·335	73·600	77·571	83·308	88·250	94·037	98·028	109·791
70	45·442	47·893	51·739	55·329	59·898	63·346	69·334	75·689	79·715	85·527	90·531	96·388	100·425	112·317

Variance ratio (or F distribution) 5% points

i.e. F_{v_1, v_2} at 5 % level

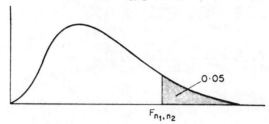

v_1 v_2	1	2	3	4	5	6	8	12	24	∞
1	161·4	199·5	215·7	224·6	230·2	234·0	238·9	243·9	249·0	254·3
2	18·51	19·00	19·16	19·25	19·30	19·33	19·37	19·41	19·45	19·50
3	10·13	9·55	9·28	9·12	9·01	8·94	8·84	8·74	8·64	8·53
4	7·71	6·94	6·59	6·39	6·26	6·16	6·04	5·91	5·77	5·63
5	6·61	5·79	5·41	5·19	5·05	4·95	4·82	4·68	4·53	4·36
6	5·99	5·14	4·76	4·53	4·39	4·28	4·15	4·00	3·84	3·67
7	5·59	4·74	4·35	4·12	3·97	3·87	3·73	3·57	3·41	3·23
8	5·32	4·46	4·07	3·84	3·69	3·58	3·44	3·28	3·12	2·93
9	5·12	4·26	3·86	3·63	3·48	3·37	3·23	3·07	2·90	2·71
10	4·96	4·10	3·71	3·48	3·33	3·22	3·07	2·91	2·74	2·54
11	4·84	3·98	3·59	3·36	3·20	3·09	2·95	2·79	2·61	2·40
12	4·75	3·88	3·49	3·26	3·11	3·00	2·85	2·69	2·50	2·30
13	4·67	3·80	3·41	3·18	3·02	2·92	2·77	2·60	2·42	2·21
14	4·60	3·74	3·34	3·11	2·96	2·85	2·70	2·53	2·35	2·13
15	4·54	3·68	3·29	3·06	2·90	2·79	2·64	2·48	2·29	2·07
16	4·49	3·63	3·24	3·01	2·85	2·74	2·59	2·42	2·24	2·01
17	4·45	3·59	3·20	2·96	2·81	2·70	2·55	2·38	2·19	1·96
18	4·41	3·55	3·16	2·93	2·77	2·66	2·51	2·34	2·15	1·92
19	4·38	3·52	3·13	2·90	2·74	2·63	2·48	2·31	2·11	1·88
20	4·35	3·49	3·10	2·87	2·71	2·60	2·45	2·28	2·08	1·84
21	4·32	3·47	3·07	2·84	2·68	2·57	2·42	2·25	2·05	1·81
22	4·30	3·44	3·05	2·82	2·66	2·55	2·40	2·23	2·03	1·78
23	4·28	3·42	3·03	2·80	2·64	2·53	2·38	2·20	2·00	1·76
24	4·26	3·40	3·01	2·78	2·62	2·51	2·36	2·18	1·98	1·73
25	4·24	3·38	2·99	2·76	2·60	2·49	2·34	2·16	1·96	1·71
26	4·22	3·37	2·98	2·74	2·59	2·47	2·32	2·15	1·95	1·69
27	4·21	3·35	2·96	2·73	2·57	2·46	2·30	2·13	1·93	1·67
28	4·20	3·34	2·95	2·71	2·56	2·44	2·29	2·12	1·91	1·65
29	4·18	3·33	2·93	2·70	2·54	2·43	2·28	2·10	1·90	1·64
30	4·17	3·32	2·92	2·69	2·53	2·42	2·27	2·09	1·89	1·62
40	4·08	3·23	2·84	2·61	2·45	2·34	2·18	2·00	1·79	1·51
60	4·00	3·15	2·76	2·52	2·37	2·25	2·10	1·92	1·70	1·39
120	3·92	3·07	2·68	2·45	2·29	2·17	2·02	1·83	1·61	1·25
∞	3·84	2·99	2·60	2·37	2·21	2·10	1·94	1·75	1·52	1·00

v (Greek nu) = number of degrees of freedom = $n_1 - 1$ (*see* p. 116)
v_2 = number of degrees of freedom = $n_2 - 1$

Lower 5 % points are found by interchange of v_1 and v_2, i.e. v_1 must always correspond with the greater mean square.

Variance ratio (or F distribution) 1% points

i.e. F_{v_1, v_2} at 1% level

v_2 \ v_1	1	2	3	4	5	6	8	12	24	∞
1	4052	4999	5403	5625	5764	5859	5982	6106	6234	6366
2	98·50	99·00	99·17	99·25	99·30	99·33	99·37	99·42	99·46	99·50
3	34·12	30·82	29·46	28·71	28·24	27·91	27·49	27·05	26·60	26·12
4	21·20	18·00	16·69	15·98	15·52	15·21	14·80	14·37	13·93	13·46
5	16·26	13·27	12·06	11·39	10·97	10·67	10·29	9·89	9·47	9·02
6	13·74	10·92	9·78	9·15	8·75	8·47	8·10	7·72	7·31	6·88
7	12·25	9·55	8·45	7·85	7·46	7·19	6·84	6·47	6·07	5·65
8	11·26	8·65	7·59	7·01	6·63	6·37	6·03	5·67	5·28	4·86
9	10·56	8·02	6·99	6·42	6·06	5·80	5·47	5·11	4·73	4·31
10	10·04	7·56	6·55	5·99	5·64	5·39	5·06	4·71	4·33	3·91
11	9·65	7·20	6·22	5·67	5·32	5·07	4·74	4·40	4·02	3·60
12	9·33	6·93	5·95	5·41	5·06	4·82	4·50	4·16	3·78	3·36
13	9·07	6·70	5·74	5·20	4·86	4·62	4·30	3·96	3·59	3·16
14	8·86	6·51	5·56	5·03	4·69	4·46	4·14	3·80	3·43	3·00
15	8·68	6·36	5·42	4·89	4·56	4·32	4·00	3·67	3·29	2·87
16	8·53	6·23	5·29	4·77	4·44	4·20	3·89	3·55	3·18	2·75
17	8·40	6·11	5·18	4·67	4·34	4·10	3·79	3·45	3·08	2·65
18	8·28	6·01	5·09	4·58	4·25	4·01	3·71	3·37	3·00	2·57
19	8·18	5·93	5·01	4·50	4·17	3·94	3·63	3·30	2·92	2·49
20	8·10	5·85	4·94	4·43	4·10	3·87	3·56	3·23	2·86	2·42
21	8·02	5·78	4·87	4·37	4·04	3·81	3·51	3·17	2·80	2·36
22	7·94	5·72	4·82	4·31	3·99	3·76	3·45	3·12	2·75	2·31
23	7·88	5·66	4·76	4·26	3·94	3 71	3·41	3·07	2·70	2·26
24	7·82	5·61	4·72	4·22	3·90	3·67	3·36	3·03	2·66	2·21
25	7·77	5·57	4·68	4·18	3·86	3·63	3·32	2·99	2·62	2·17
26	7·72	5·53	4·64	4·14	3·82	3·59	3·29	2·96	2·58	2·13
27	7·86	5·49	4·60	4·11	3·78	3·56	3·26	2·93	2·55	2·10
28	7·64	5·45	4·57	4·07	3·75	3·53	3·23	2·90	2·52	2·06
29	7·60	5·42	4·54	4·04	3·73	3·50	3·20	2·87	2·49	2·03
30	7·56	5·39	4·51	4·02	3·70	3·47	3·17	2·84	2·47	2·01
40	7·31	5·18	4·31	3·83	3·51	3·29	2·99	2·66	2·29	1·80
60	7·08	4·98	4·13	3·65	3·34	3·12	2·82	2·50	2·12	1·60
120	6·85	4·79	3·95	3·48	3·17	2·96	2·66	2·34	1·95	1·38
∞	6·64	4·60	3·78	3·32	3·02	2·80	2·51	2·18	1·79	1·00

v_1 = number of degrees of freedom = $n_1 - 1$ (*see* p. 116)
v_2 = number of degrees of freedom = $n_2 - 1$

Lower 1 % points are found by interchange of v_1 and v_2, i.e. v_1 must always correspond with the greater mean square.

Random numbers

03 47 43 73 86	36 96 47 36 61	46 98 63 71 62	33 26 16 80 45	60 11 14 10 95
97 74 24 67 62	42 81 14 57 20	42 53 32 37 32	27 07 36 07 51	24 51 79 89 73
16 76 62 27 66	56 50 26 71 07	32 90 79 78 53	13 55 38 58 59	88 97 54 14 10
12 56 85 99 26	96 96 68 27 31	05 03 72 93 15	57 12 10 14 21	88 26 49 81 76
55 59 56 35 64	38 54 82 46 22	31 62 43 09 90	06 18 44 32 53	23 83 01 30 30
16 22 77 94 39	49 54 43 54 82	17 37 93 23 78	87 35 20 96 43	84 26 34 91 64
84 42 17 53 31	57 24 55 06 88	77 04 74 47 67	21 76 33 50 25	83 92 12 06 76
63 01 63 78 59	16 95 55 67 19	98 10 50 71 75	12 86 73 58 07	44 39 52 38 79
33 21 12 34 29	78 64 56 07 82	52 42 07 44 38	15 51 00 13 42	99 66 02 79 54
57 60 86 32 44	09 47 27 96 54	49 17 46 09 62	90 52 84 77 27	08 02 73 43 28
18 18 07 92 46	44 17 16 58 09	79 83 86 19 62	06 76 50 03 10	55 23 64 05 05
26 62 38 97 75	84 16 07 44 99	83 11 46 32 24	20 14 85 88 45	10 93 72 88 71
23 42 40 64 74	82 97 77 77 81	07 45 32 14 08	32 98 94 07 72	93 85 79 10 75
52 36 28 19 95	50 92 26 11 97	00 56 76 31 38	80 22 02 53 53	86 60 42 04 53
37 85 94 35 12	83 39 50 08 30	42 34 07 96 88	54 42 06 87 98	35 85 29 48 39
70 29 17 12 13	40 33 20 38 26	13 89 51 03 74	17 76 37 13 04	07 74 21 19 30
56 62 18 37 35	96 83 50 87 75	97 12 25 93 47	70 33 24 03 54	97 77 46 44 80
99 49 57 22 77	88 42 95 45 72	16 64 36 16 00	04 43 18 66 79	94 77 24 21 90
16 08 15 04 72	33 27 14 34 09	45 59 34 68 49	12 72 07 34 45	99 27 72 95 14
31 16 93 32 43	50 27 89 87 19	20 15 37 00 49	52 85 66 60 44	36 68 88 11 80
68 34 30 13 70	55 74 30 77 40	44 22 78 84 26	04 33 46 09 52	68 07 97 06 57
74 57 25 65 76	59 29 97 68 60	71 91 38 67 54	13 58 18 24 76	15 54 55 95 52
27 42 37 86 53	48 55 90 65 72	96 57 69 36 10	96 46 92 42 45	97 60 49 04 91
00 39 68 29 61	66 37 32 20 30	77 84 57 03 29	10 45 65 04 26	11 04 96 67 24
29 94 98 94 24	68 49 69 10 82	53 75 91 93 30	34 25 20 57 27	40 48 73 51 92
16 90 82 66 59	83 62 64 11 12	67 19 00 71 74	60 47 21 29 68	02 02 37 03 31
11 27 94 75 06	06 09 19 74 66	02 94 37 34 02	76 70 90 30 86	38 45 94 30 38
35 24 10 16 20	33 32 51 26 38	79 78 45 04 91	16 92 53 56 16	02 75 50 95 98
38 23 16 86 38	42 38 97 01 50	87 75 66 81 41	40 01 74 91 62	48 51 84 08 32
31 96 25 91 47	96 44 33 49 13	34 86 82 53 91	00 52 43 48 85	27 55 26 89 62
66 67 40 67 14	64 05 71 95 86	11 05 65 09 68	76 83 20 37 90	57 16 00 11 66
14 90 84 45 11	75 73 88 05 90	52 27 41 14 86	22 98 12 22 08	07 52 74 95 80
68 05 51 18 00	33 96 02 75 19	07 60 62 93 55	59 33 82 43 90	49 37 38 44 59
20 46 78 73 90	97 51 40 14 02	04 02 33 31 08	39 54 16 49 36	47 95 93 13 30
64 19 58 97 79	15 06 15 93 20	01 90 10 75 06	40 78 78 89 62	02 67 74 17 33
05 26 93 70 60	22 35 85 15 13	92 03 51 59 77	59 56 78 06 83	52 91 05 70 74
07 97 10 88 23	09 98 42 99 64	61 71 62 99 15	06 51 29 16 93	58 05 77 09 51
68 71 86 85 85	54 87 66 47 54	73 32 08 11 12	44 95 92 63 16	29 56 24 29 48
26 99 61 65 53	58 37 78 80 70	42 10 50 67 42	32 17 55 85 74	94 44 67 16 94
14 65 52 68 75	87 59 36 22 41	26 78 63 06 55	13 08 27 01 50	15 29 39 39 43
17 53 77 58 71	71 41 61 50 72	12 41 94 96 26	44 95 27 36 99	02 96 74 30 83
90 26 59 21 19	23 52 23 33 12	96 93 02 18 39	07 02 18 36 07	25 99 32 70 23
41 23 52 55 99	31 04 49 69 96	10 47 48 45 88	13 41 43 89 20	97 17 14 49 17
60 20 50 81 69	31 99 73 68 68	35 81 33 03 76	24 30 12 48 60	18 99 10 72 34
91 25 38 05 90	94 58 28 41 36	45 37 59 03 09	90 35 57 29 12	82 62 54 65 60
35 50 57 74 37	98 80 33 00 91	09 77 93 19 82	74 94 80 04 04	45 07 31 66 49
85 22 04 39 43	73 81 53 94 79	33 62 46 86 28	08 31 54 46 31	53 94 13 38 47
09 79 13 77 48	73 82 97 22 21	05 03 27 24 83	72 89 44 05 60	35 80 39 94 88
88 75 80 18 14	22 95 75 42 49	39 32 82 22 49	02 48 07 70 37	16 04 61 67 87
90 96 23 70 00	39 00 03 06 90	55 85 78 38 36	94 37 30 69 32	90 89 00 76 33

References

General references which cover most of the material in the book but not necessarily in a marketing context:

1. CROFT, D., *Applied Statistics for Management Studies* (Macdonald and Evans, 1969).
2. CROXTON, F. E. and COWDEN, D. J., *Applied General Statistics* (Prentice-Hall/Pitman, 2nd Edition, 1960).
3. FREUND, S. E. and WILLIAMS, F. J., *Elementary Business Statistics* (Prentice-Hall, 1964).
4. HOEL, P. G., *Elementary Statistics* (Wiley, 1960).
5. LEWIS, T. W. and FOX, R. A., *Managing with Statistics* (Oliver and Boyd, 1969).
6. MORONEY, M. J., *Facts from Figures* (Pelican, 3rd Edition, 1964).
7. SPIEGEL, M. R., *Theory and Problems of Statistics* (Schaum Publishing Company, 1961).
   A more rigorous treatment of statistics than that given in 1–7 is
8. WALPOLE, R. E., *Introduction to Statistics* (Collier-Macmillan, 1968).

Relevant journals include:

9. *British Journal of Marketing.*
10. Industrial Marketing Research Association *Journal.*
11. *Journal of Advertising Research.*
12. *Journal of Marketing* (American Marketing Association).
13. *Journal of Marketing Research* (American Marketing Association).
14. *Journal of the Market Research Society.*
15. *Marketing* (Journal of the Institute of Marketing).
16. *The Statistician* (Journal of the Institute of Statisticians).

Additional references related to specific chapters:

CHAPTER 3

1. LINDLEY, D. V., *Introduction to Probability and Statistics, from 'A Bayesian Viewpoint'* (Cambridge University Press, 1965).
2. SAVAGE, L. J., *The Foundations of Statistical Inference* (Methuen, 1962).

CHAPTER 4
1. LOWE, C. W. *Industrial Statistics* Vol. 1 (Business Books, 1968).
2. BARTHOLOMEW, D. J. and BASSETT, E. E., *Let's Look at Figures* (Penguin, 1971).

CHAPTER 5
1. LOWE, C. W., *Industrial Statistics* Vol. 1 (Business Books, 1968).
2. DEMING, W. E., *Some Theory of Sampling* (Dover, 1966).

CHAPTER 6
1. DUCKWORTH, W. E., *Statistical Techniques in Technological Research: An Aid to Research Productivity* (Methuen, 1967).
2. LANGLEY, R., *Practical Statistics* (Pan Piper, 1970).

CHAPTER 7
1. DUCKWORTH, W. E., *Statistical Techniques in Technological Research: An Aid to Research Productivity* (Methuen, 1968).
2. GIBBONS, J. D., *Non-Parametric Statistical Inference* (McGraw-Hill, 1971).
3. LANGLEY, R., *Practical Statistics* (Pan Piper, 1970).
4. QUENOUILLE, M. H., *Rapid Statistical Calculations* (Griffin, 1971).

CHAPTER 8
1. AITCHISON, J., *Solving Problems in Statistics II* (Oliver and Boyd, 1971).

CHAPTER 9
1. YAMANE, T., *Statistics: An Introductory Analysis* (Harper and Row, 1964).
2. THIRKETTLE, G. L., *Wheldon's Business Statistics and Statistical Method* (Macdonald and Evans, 1968).

CHAPTER 11
1. BATTERSBY, A., *Sales Forecasting* (Cassell, 1968).
2. BATTY, M. 'Monitoring an exponential smoothing forecasting system' (*Operational Research Quarterly*, Vol. 20, No. 3, 1969).
3. BROWN, R. G., *Smoothing, Forecasting and Prediction of Discrete Time Series* (Prentice-Hall, 1962).
4. TRIGG, D. W., 'Monitoring a forecasting system' (*Operational Research Quarterly*, Vol. 15, No. 3, 1964).

5. TRIGG, D. W. and LEACH, A. G., 'Exponential smoothing with an adaptive response rate' (*Operational Research Quarterly*, Vol. 18, No. 1, 1967).
6. *Cumulative Sum Techniques I.C.I. Monograph* No. 3 (Oliver and Boyd, 1964).
7. *Short Term Forecasting, I.C.I. Monograph* No. 2 (Oliver and Boyd, 1964).
8. LEWIS, C. D., *Industrial Forecasting Techniques* (Machinery, 1970).

CHAPTER 12

1. COCHRANE, W. G. and COX, G. M., *Experimental Designs* (Wiley, 1957).
2. COOLEY, W. W. and LOHNES, P. R., *Multivariate Data Analysis* (Wiley, 1971).
3. COX, D. R., *Planning of Experiments* (Wiley, 1966).
4. DAVIES, O. L. (Editor) *Statistical Methods in Research and Production* (Oliver and Boyd, 1957).
5. DUCKWORTH, W. E., *Statistical Techniques in Technological Research: An Aid to Research Productivity* (Methuen, 1966).
6. FRANK, R. E. and GREEN, P. E., *Quantitative Methods in Marketing* (Prentice-Hall, 1967).
7. MOSER, C. A. and KALTON, G., *Survey Methods in Social Investigation* (Heinemann, 1971).
8. OPPENHEIM, A. N., *Questionnaire Design and Attitude Measurement* (Heinemann, 1966).
9. RALSTON, A. and WILF, H. S., *Mathematical Methods for Digital Computers, Vol. I* (Wiley, 1960).
10. SMILLIE, K. W., *An Introduction to Regression and Correlation* (Academic Press, 1966).
11. STEPHAN, F. F. and MCCARTHY, P. J., *Sampling Opinions* (Wiley, 1963).

Answers to Exercises

CHAPTER 1

1. (a) 11   (b) 46
2. 50
3. (a) Yes   (b) Yes   (c) No

CHAPTER 2

3. mean = 8·05 packets.
5. (a) modal volume represented by class 9–10 cm³ (or by class mark of 9·5 cm³).
   (b) median value is 9·5 cm³ which is the class mark of the class 9–10 cm³ (by interpolation the median = 9·5 cm³).
9. mean = 25·4 tonnes, median = 25 tonnes, mode = 20·9 tonnes.
10. mean = 15·0, median = 15, mode = 15.
11. mean = 14·2 grams/litre, standard deviation = 0·805 grams, coefficient of variation = 0·057.
12. mean = 99·1, median = 94·5 (i.e. class mark of class 90–99), standard deviation = 16·5.
13. variance = 3·8.
14. mean = 44·9 years. Interquartile deviation represented by age of employee 982 minus age of employee 327 divided by 2, i.e. (range (41–46) — range (31–36))/2 or (class mark 43·5 — class mark 33·5)/2 = 10·0/2 = 5 years.
15. By interpolation median = 34·0 years and the quartiles are 27·5, 34·0, and 40·25 years.
16. median represented by class interval (4–6).
17. The median = 34·2 years. The quartiles are 17·5, 34·2, and 51 years. The interquartile range is (51−17·5) = 33·5 years, therefore the semi-interquartile range = 16·75 years. Percentages are: 12·5, 12·5, 14·5.
18. Quartile coefficient of skewness = 0·024.

CHAPTER 3

1. 29·2%
2. 0·595

3. 0·4772
4. (a) 39·06%   (b) 20·23%
5. 81·12%
6. (a) 3; 15; 1; 120; 120
   (Note that $^nC_r = {}^nC_{n-r}$)
   (b) $1 + 3 + 3 + 1 = 8 = 2^3$
   $1 + 5 + 10 + 10 + 5 + 1 = 32 = 2^5$

   (Note that $\sum_{r=0}^{n} {}^nC_r = 2^n$. This sometimes enables useful arithmetic check.)
7. $p = 0·2$ $q = 0·8$ $P(0 \text{ in } 5) = {}^5C_0 \times 0·2^0 \times 0·8^5 = 0·32768$

No. of discriminating housewives:	0	1	2	3	4	5
Probability (binomial distribution)	0·3277	0·4096	0·2048	0·0512	0·0064	0·0003
No. of groups	328	410	205	51	6	(0·3)

8. $np = 1 = \mu$ $\sqrt{npq} = \sqrt{0·8} = 0·894 = \sigma$

No. of discriminating housewives:	0	1	2	3	4	5
Probability	0·241	0·425	0·241	0·044	0·003	~0

Agreement is poor because n is small and p is *not* close to 0·5. $np \ll 5·0$.

CHAPTER 4

1. (a) 1·41; (b) 0·71; (c) 0·32
2. (a) 1·40; (b) 0·68; (c) 0·245
3. Machine settings should be (a) 254·24 grams; (b) 248·85 grams; and
   (c) 249·62 grams.
4. (a) Sample mean = 61·0; sample size = 4.
   Because population is normally distributed the means of even very small samples will be normally distributed. If they come from population $\mu = 60$, $\sigma = 10$ this sample mean distribution will have mean 60 and S.E. = $10/\sqrt{n}$. In this case S.E. = 5. Using Normal Distribution tables the problem of getting a value as much as $\dfrac{61-60}{5} = 0·2$ units distinct from the true mean is 0·8414.

   (b) Sample mean = 61·4 $n = 9$ S.E. = 3·33
   $$\frac{61·4 - 60·0}{3·33} = \frac{1·4}{3·33} = 0·42$$
   Associated probability = 0·6744
   (c) Sample mean = 66·66 $n = 9$ S.E. = 3·33
   $$\frac{66·66 - 60·00}{3·33} = 2·0$$
   Associated probability = 0·0456

5. (a) 99% limits $= 28\cdot27 \pm 2\cdot58 \times 0\cdot47$

   (b) Estimated population variance $= \dfrac{5\cdot28 \times 100}{99} = 5\cdot33$

      Estimated standard error of sample means $= \sqrt{\dfrac{5\cdot33}{100}} = 0\cdot231$

      95% confidence limits $= 28\cdot27 \pm 1\cdot96 \times 0\cdot231$

      99%   ,,   ,, $= 28\cdot27 \pm 2\cdot58 \times 0\cdot231$

6. (a) Estimated population $= \dfrac{8}{40} = 0\cdot20$

      Standard error of estimate $= \sqrt{\dfrac{0\cdot20 \times 0\cdot80}{40}} = 0\cdot063$

      95% limits $= 0\cdot20 \pm 1\cdot96 \times 0\cdot063 = 0\cdot08 - 0\cdot32$

   (b) Estimated population $= \dfrac{800}{4,000} = 0\cdot20$

      Standard error of estimate $= \sqrt{\dfrac{0\cdot20 \times 0\cdot80}{4,000}} = 0\cdot0063$

      95% limits $= 0\cdot20 \pm 1\cdot96 \times 0\cdot0063 = 0\cdot188 - 0\cdot212$

CHAPTER 5

1. $n = 97$
2. (a) $n = 3,160$; (b) $n = 1,580$; (c) $n = 2,236$
3. (a) $n = 6,145$; (b) $n = 3,458$; (c) $n = 9,604$
   In general the nearer to $0\cdot5$ the required proportion the larger the sample required to estimate it to any required degree of precision.
4. Sample partitions into $3:8:15:23\cdot5$

CHAPTER 6

1. No significant effect at 5% level
   Chi-squared value $= 2\cdot51$
   Degrees of freedom $= 2$
2. (a) F value $= 1\cdot0$ hence combination of samples is permissible.
     $t = 0\cdot94$; 14 d of f. Not significant at 5% level
   (b) Restaurant A: $t = 1\cdot66$; 7 d of f. Not significant at 5% level
     Restaurant B: $t = 0\cdot33$; 7 d of f. Not significant at 5% level
   (c) Assumption of a normal distribution of g solid/l. in hypothesized parent population is necessary. The design of experiment is also dubious because the second restaurant, could at least become aware that an inspection was imminent and reset accordingly!

3. F value $= 1\cdot3$ with 9 and 7 d of f. Not significant at 2% level therefore sample variances can be combined.

$t = 4\cdot7$; 16 d of f. Significant at $0\cdot1$% level.

4. $t = 2\cdot57$; d of $f = 8$

5. $t = 1\cdot62$; d of $f = 16$. Not significant at 5% level.

6. $\chi^2 = 0\cdot99$; d of $f = 3$

CHAPTER 7

1. (a) $z = 0\cdot92$. Associated Normal distribution probability $= 0\cdot3576$. Samples may be considered as coming from identically distributed populations.

(b) $z = 2\cdot63$ Associated Normal distribution probability $= 0\cdot0086$. Significant at 1% level.

2. (a) $z = 1\cdot52$. Associated Normal distribution probability $= 0\cdot1286$. Not significant at 5% level.

(b) $z = 2\cdot70$. Associated Normal distribution probability $= 0\cdot0070$. Significant at 1% level.

3. $z = 1\cdot39$. Associated Normal distribution probability $= 0\cdot1646$. There is no evidence of a significant increase in sales.

4. $\chi^2 = 0\cdot15$. No significant difference amongst cities.

5. The test statistic is $(12-4)/2\sqrt{16} = 1\cdot0$. By this test the differences are just significant at the 5% level.

6. The test statistics are:

1(a) 4; 1(b) 6; 2(a) 5; 2(b) 6

7. In both cases the test statistic is $9/2\sqrt{11} = 1\cdot35$.

Both correlations are significant at the 5% level.

CHAPTER 8

1. (a) $20/300 = 0\cdot67$

(b) $150/2000 = 0\cdot075$

(c) $550/1280 = 0\cdot43$

(d) $1080/1680 = 0\cdot64$

2. $P(n \mid 2h) = 0\cdot961 = P'(n) \ P(dh) \mid 2h) = 0\cdot039 = P'(dh)$

$P(n \mid 3h) = P(n \mid 2h)$ modified by h in next trial

$$= \frac{P(1h \mid n) \times P'(n)}{P(1h \mid n) \times P'(n) + P(1h \mid dh) \times P'(dh)}$$

$$= \frac{0\cdot5 \times 0\cdot961}{0\cdot5 \times 0\cdot961 + 1\cdot00 \times 0\cdot039} = 0\cdot925$$

3. Application of the test should result in 70·4% of the successful candidates being 'top-class' salesmen.

4. (a) The minimum expected cost sample size is 3. The associated cost is (approximately) £39·01.

   (b) The minimum expected cost sample size is 1. The associated cost is (approximately) £39·50.

CHAPTER 9

1. 123

2. $I_{72} = 100$, $I_{73} = 110$, $I_{74} = 107$, $I_{75} = 119·1$

3. (a) 114·0 ∴ claim disallowed

   (b) 112·7 ∴ not affected

4. $I_{57} = 104·6$, $I_{58} = 104·0$, $I_{59} = 106·6$

   $I_{60} = 108·3$, $I_{61} = 109·2$, $I_{62} = 110·0$

5. $I = \dfrac{\Sigma \dfrac{p_1}{p_0} \cdot v_0}{\Sigma v_0} = \dfrac{\Sigma \dfrac{p_1}{p_0} \cdot p_0 q_0}{\Sigma p_0 q_0} = \dfrac{\Sigma p_1 q_0}{\Sigma p_0 q_0} = 114$

6. Value of nitrogenous in 1968 = £16·1/tonne ∴ 1978 quantity value is £1,016.

   Value of nitrogenous in 1978 = £42·3/tonne ∴ 1968 quantity value is £1,024.

   Value of other in 1968 = £4·0/tonne ∴ 1978 quantity value is £2,945.

   Value of other in 1978 = £10·77/ton ∴ 1968 quantity value is £4,945.

   Value $I_{78,68} = 268·5$

   Value $I_{68,78} = 37·4$

   Volume $I_{78,68} = \dfrac{\Sigma q_{78} p_{68}}{\Sigma q_{68} p_{68}} = 178·3$

   Volume $I_{68,78} = 56·3$

7. $I_{76} = 152$; $I_{77} = 170$

CHAPTER 10

1. Weight $= 186·4 + 8·2$ weeks

2. Product-moment correlation coefficient $= 0·72$

   Rank correlation coefficient $= 0·8$

3. $r = 0·463$, $t = 1·38$, $t_7{}^{5\%} = 2·365$, not significant.

4. $\rho = -0·14$, $t = 0·34$, not significant

5. Rank correlation coefficient $= 0·585$. $t = 2·04$, $t_8{}^{5\%} = 2·306$, not significant. Therefore share price is not related to F.T. index.

6. Correlation coefficient $= -0·76$ which is significant at 1% level.

   TV licences $= 474·72 - 42·34$ cinema admissions.

7. Correlation coefficient $= 0·73$ which is significant at 5% level.

   $78 \pm 23·7$. No.

CHAPTER 11

1.

$i =$	1	2	3	4	5	6	7	8
(a)	9·5	9·75	9·25	9·25	8·75	9·00	9·25	8·75
(b)	9·4	9·5	9·0	9·0	9·0	9·4	9·1	8·7
(c)	10·6	10·1	7·6	8·7	8·9	10·6	8·5	7·3

2. Forecast for month 19 is 26·66
   Forecast for month 23 is 28·06
3. Forecast for month 31 is 12·13
   Forecast for month 34 is 13·18
4. Forecast for 13th quarter is 7·56
   Forecast for 14th quarter is 5·42
5. 8·5
6.

	Quarter	1	2	3	4
Deviations	Additive	−0·20	−0·16	0·16	0·20
	Ratio	0·87	0·89	1·11	1·13

7. 33

Index

Note: Entries consisting of Greek letters or mathematical notations are listed at the end of the index.

ECONOMICS:
An Introduction for Students of Business and Marketing
Frank Livesey

Here at last is an economics text grounded on reality. The author believes that producers are at the heart of the economics system, and he discusses both the way their activities — including the marketing process — affect the community and the way they respond to the community's needs.

PEOPLE AND PERFORMANCE
The Best of Peter Drucker on Management
Peter F. Drucker

Business management executives and students will find within these pages a survey of developments central to management thought and which provide the basis for the understanding of current practice.

MANAGING IN TURBULENT TIMES
Peter F. Drucker

A coruscating survey of up-to-the-minute social and economic issues. The world's greatest management thinker provides unique and sometimes startling insights into, for example, the potentially dramatic effects of new technologies; the steady change from an essentially 'manual-labour force' to a 'knowledge force'; the effects of demographic changes, and much else. This book deserves to be read by a wide audience, but especially by all grades of management.

QUANTITATIVE METHODS FOR BUSINESS STUDENTS
Roger Carter

This simple exposition of all aspects of numerical techniques, ranging from fractions, decimals and elementary probability to calculus, expectation and queuing theory, has been designed specifically for students without a strong mathematical background. It has a number of special features, most important of which are the exercises as well as practical assignments at the end of each chapter and the solutions section at the end of the book.